Approach your exams the IB way

Geography

IB DIPLOMA PROGRAMME

IB PREPARED

AF795440

Briony Cooke
Garrett Nagle

International Baccalaureate
Baccalauréat International
Bachillerato Internacional

IB Prepared
Approach your exams the IB way
Geography SL and HL

Published May 2012

Published on behalf of the International Baccalaureate Organization, a not-for-profit educational foundation of 15 Route des Morillons, 1218 Le Grand-Saconnex, Geneva, Switzerland by the

International Baccalaureate Organization (UK) Ltd
Peterson House, Malthouse Avenue, Cardiff Gate
Cardiff, Wales CF23 8GL
United Kingdom
Phone: +44 29 2054 7777
Fax: +44 29 2054 7778
Website: www.ibo.org

IB merchandise and publications can be purchased through the IB store at http://store.ibo.org. General ordering queries should be directed to the Sales and Marketing Department in Cardiff.

Phone: +44 29 2054 7746
Fax: +44 29 2054 7779
Email: sales@ibo.org

British Library Cataloguing in Publication Data.
A catalogue record for this book is available from the British Library.

ISBN: 978-1-906345-55-6

Cover design by Pentacorbig
Typeset by Wearset Ltd
Printed and bound by EPC Direct

Item code 3114

2016 2015 2014 2013 2012
10 9 8 7 6 5 4 3 2 1

Acknowledgments

Thanks to Richard Allaway and Kathleen Noreisch for their advice on the IB Diploma Programme geography.

Dedications

To Henry for his support and dedication to domestic duty during the writing of this book
and to Angela, Rosie, Patrick and Bethany – for their continued support and good humour

IB learner profile

The aim of all IB programmes is to develop internationally minded people who, recognizing their common humanity and shared guardianship of the planet, help to create a better and more peaceful world.

IB learners strive to be:

Inquirers	They develop their natural curiosity. They acquire the skills necessary to conduct inquiry and research and show independence in learning. They actively enjoy learning and this love of learning will be sustained throughout their lives.
Knowledgeable	They explore concepts, ideas and issues that have local and global significance. In so doing, they acquire in-depth knowledge and develop understanding across a broad and balanced range of disciplines.
Thinkers	They exercise initiative in applying thinking skills critically and creatively to recognize and approach complex problems, and make reasoned, ethical decisions.
Communicators	They understand and express ideas and information confidently and creatively in more than one language and in a variety of modes of communication. They work effectively and willingly in collaboration with others.
Principled	They act with integrity and honesty, with a strong sense of fairness, justice and respect for the dignity of the individual, groups and communities. They take responsibility for their own actions and the consequences that accompany them.
Open-minded	They understand and appreciate their own cultures and personal histories, and are open to the perspectives, values and traditions of other individuals and communities. They are accustomed to seeking and evaluating a range of points of view, and are willing to grow from the experience.
Caring	They show empathy, compassion and respect towards the needs and feelings of others. They have a personal commitment to service, and act to make a positive difference to the lives of others and to the environment.
Risk-takers	They approach unfamiliar situations and uncertainty with courage and forethought, and have the independence of spirit to explore new roles, ideas and strategies. They are brave and articulate in defending their beliefs.
Balanced	They understand the importance of intellectual, physical and emotional balance to achieve personal well-being for themselves and others.
Reflective	They give thoughtful consideration to their own learning and experience. They are able to assess and understand their strengths and limitations in order to support their learning and personal development.

Table of contents

1. Introduction

As an IB student, you are provided with many resources on the path to your final geography standard level (SL) or higher level (HL) exams. This book is just one of the resources to help you prepare.

How to use this book

Welcome to *IB Prepared*! The main purpose of this book is to help you, the student, prepare for your exams. Using actual student work from past exams, this book provides practical advice on how to approach exam questions, explaining how marks are given, how marks may be lost, as well as explaining what the examiners are looking for in the questions. The commentaries are provided by experienced IB teachers and examiners, so what you read is based on the knowledge of exam setting, marking and teaching of IB geography.

This book contains helpful advice in preparing for IB exams. However, it cannot replace the content material that you have been taught, and are expected to have learned, before taking your exam. Your teachers are a major source of support and guidance, and this book is supposed to supplement your classroom experiences, not replace them. We suggest that, as you read through this book, you work through the past exam questions first, before reading the student answers and the examiners' commentaries. In this way you will play an active role in your revision, get more out of the book and become more aware of the demands of the exams.

What is in this book?

- **Chapter 2** explains the meaning of command terms. It illustrates the level of complexity associated with certain command terms (which are classified according to four assessment objectives—AOs) and illustrates what is required with a selected group of command terms. It also looks at some of the specific terminology associated with IB geography.

- **Chapter 3** focuses on the structure of the exams and distinguishes between the needs for SL and HL, examining the assessment criteria for each of the papers.

- **Chapter 4** provides some top tips for exam preparation and during the exam.

- **Chapter 5** looks at an essential area of the exams—stimulus material—and offers advice on how to work with this type of material, examining how to respond to data stimulus-style questions.

- The rest of the book has been broken up to match the IB syllabus: **chapter 6** deals with the core theme; chapters **7–13** examine the optional themes; and **chapter 14** focuses on global interactions.

    - The first part of each chapter examines the key command terms and objectives that you would have come across in class, and which are outlined in the subject guide. This provides a brief reminder of each sub-topic in the guide. Key definitions are provided throughout, but these are only a selection. You should also refer to the definitions in the *Geography guide*.

    - The second part of these chapters prepares you for the types of question that you may encounter. Past paper questions are used, covering all areas of the syllabus, accompanied by guidance on how to approach the exam question followed by examples of high-scoring (not necessarily model answers), medium-quality and low-quality answers. To help you gain an insight into the way an examiner works, for each answer we provide you with the mark that was given and an explanation of the weaknesses and strengths. There are many ways to answer geography questions and much will depend on the case studies you use to support your answers; different case studies could result in very different answers. At the end of the extended response answers we provide an examiner report relating the answer to the assessment criteria.

- In **chapter 15** you will find further examples of exam questions, allowing you to see exactly what to expect during the exams and enabling you to **put into practice** what you have learned from this book.

Please remember that simply reading this book will not guarantee you a high mark. It is a good understanding of geographical knowledge and how it can be applied during the exam that will help you to achieve success.

We hope that you find the rest of this book useful and we wish you good luck and every success in your geography exams.

2. The command terms

In this chapter you will gain an understanding of command terms and how they relate to the exam questions.

Command terms are the words in the exam question itself that tell you how to approach the question, and the detail and the depth expected. It is crucial that you interpret these terms correctly.

Your exam papers are designed to test four assessment objectives (AOs): AO1, AO2, AO3 and AO4, and all the command terms are classified accordingly.

No reference is made to these assessment objectives in any part of the exam papers. Therefore, it is important to understand their meaning and their importance in advance of the exam.

Assessment objectives and their use

Assessment objectives	What do these assess?	How are these objectives achieved in the exam?
AO1	Knowledge and understanding of specified content	By recalling knowledge of the subject content.
AO2	Application and analysis of knowledge and understanding	By recalling and bringing together knowledge, concepts and theories previously learned, and applying them to the question. By identifying and interpreting geographic patterns, processes in unfamiliar information, data and cartographic material.
AO3	Synthesis and evaluation	By examining and assessing geographic theories in the context of the question. By using geographic concepts and examples. By producing a well-structured argument using appropriate terminology. By synthesizing information, which means drawing together information and arriving at a conclusion. (HL only, in the HL extension)
AO4	Selection, use and application of a variety of appropriate skills and techniques	By selecting and applying techniques and skills using appropriate terminology.

The command term provides you with information about the question and what is required.

Before you answer a question, you should:

- underline its command terms
- look at the mark weighting and the time allowed to answer in the required depth.

The following table contains an example of how each type of command term might be used in an exam, and advice is given as to what you should cover.

Term	**Assessment objective**	**Definition**	**Sample question**	**What you should cover**
Analyse	AO2	Break down in order to bring out the essential elements or structure.	Analyse the population structure of country X. (Shown in a population pyramid.)	Describe and explain the age and gender composition of the country's population.
Compare and contrast	AO3	Give an account of similarities and differences between two (or more) items or situations, referring to both (all) of them throughout.	Compare and contrast the impacts of two recent flood events.	For the two events you should: • describe and explain the similarities (compare) in flood impact • describe and explain the differences (contrast) • conclude by reviewing the relative total impact of both flooding events.
Define	AO1	Give the precise meaning of a word, phrase, concept or physical quantity.	Define the term "agribusiness".	One mark or two marks may be allocated to this question. Look at the mark weighting to determine the detail required.
Describe	AO1	Give a detailed account.	Describe the relationship between fertilizer use and crop yield shown on the graph.	Look at the mark weighting to determine the detail required. There is no need to give reasons in your answer.
Determine	AO1	Obtain the only possible answer.	Determine the year when the rate of natural increase was highest.	Calculators are not allowed in the exam, and you may need to determine the value by eye or using a ruler.
Discuss	AO3	Offer a considered and balanced review that includes a range of arguments, factors or hypotheses. Opinions or conclusions should be presented clearly and supported by appropriate evidence.	Discuss the neo-Malthusian and anti-Malthusian views of the relationship between population and resources.	Describe and explain each of these views on the relationship between population and resources. Evaluate each view by drawing on case study evidence. Arrive at a conclusion that addresses both viewpoints but favours one more than the other.

Term	Assessment objective	Definition	Sample question	What you should cover
Distinguish	AO2	Make clear the differences between two or more concepts or items.	Distinguish between urbanization and urban sprawl.	Your answer should consist of more than two descriptions and it is essential that you emphasize the differences between them.
Draw	AO4	Represent by means of a labelled, accurate diagram or graph, using a pencil. A ruler (straight edge) should be used for straight lines. Diagrams should be drawn to scale. Graphs should have points correctly plotted (if appropriate) and joined in a straight line or smooth curve.	Draw a labelled diagram showing the principal features of glacier A in the photograph.	Be prepared to draw in all three exam papers. Your exam papers will be scanned in black and white. You should make sure that your diagrams are very distinct by drawing in black ink or pencil. Labels are brief and do not involve description. Refer to chapter 5, "Data response skills".
Estimate	AO1	Obtain an approximate value.	Estimate the area covered by volcanic ash fall.	You will have to use both the scale on the map and your ruler to obtain an approximate value.
Evaluate	AO3	Make an appraisal by weighing up the strengths and limitations.	Evaluate the strategies adopted to protect a named coastline from the impact of tropical cyclones.	Your evaluation is not just an opinion; you must provide evidence such as examples, case studies or facts to support it.
Examine	AO3	Consider an argument or a concept in a way that uncovers the assumptions and interrelationships of the issue.	Examine the relationship between social inequality and crime in cities.	This means explaining thoroughly. It requires you to demonstrate in-depth understanding and it carries high marks.
Explain	AO2	Give a detailed account including reasons or causes.	Explain the causes and consequences of the multiplier effect.	Assume that you will need to briefly describe before you explain, unless that was required in the preceding part of the question.
Identify	AO1	Provide an answer from a number of possibilities.	Identify three factors on the map that have influenced the pattern of settlement.	Only brief answers (sometimes one word) are required here.
Justify	AO3	Give valid reasons or evidence to support an answer or conclusion.	Referring to the table, select three countries with the highest levels of vulnerability to tectonic hazards. Justify your answer.	You would need to describe the factors affecting vulnerability that influenced your choice.

Term	Assessment objective	Definition	Sample question	What you should cover
Label	AO4	Add labels to a diagram.	Label features A, B and C shown in the photograph.	Each label is likely to be just one or two words. Do not describe or explain the feature.
Outline	AO1	Give a brief account or summary.	Outline the characteristics of the informal economy.	This command carries few marks, so your answer should consist of brief statements.
State	AO1	Give a specific name, value or other brief answer without explanation or calculation.	Referring to the table, state the three countries with the highest level of internet use.	Only brief answers (sometimes one word). Three words would be needed here.
Suggest	AO2	Propose a solution, hypothesis or other possible answer.	Suggest reasons for the high level of agricultural efficiency in country X.	This term is used when there are several possible answers and you may have to give reasons or a judgment.
To what extent	AO3	Consider the merits or otherwise of an argument or concept. Opinions and conclusions should be presented clearly and supported with empirical evidence and sound argument.	To what extent is a high fertility rate an advantage rather than a problem?	Your answer should consider the advantages and disadvantages of high fertility with a conclusion that favours one or the other.

The following question is analysed to show how an evaluative question might be approached.

Evaluate the success of one national population policy.

China's one child policy was introduced in 1979. Its purpose was to curb high birth rates and improve access to food and other resources, to reduce unemployment and to raise the national standard of living. The policy was hard to implement, especially in rural communities, and there were reports of enforced abortions, female infanticide and abandonment.

This paragraph is mostly descriptive and provides background to the policy.

The policy, which is still in place, has been successful in terms of achieving its aims and has reduced China's population growth rate from 1.3% in 1979 to 0.5% in 2009. This demographic change has been accompanied by dramatic economic development, with annual GDP growth rates increasing from 0% in 1979 to almost 10% in 2010. Other social changes during this period have included rapid urbanization and improvements in standard of living and literacy rates.

Positive evaluation. Demographic stability and economic growth.

However, there are now concerns about fertility rates, which have fallen to 1.4% in 2010, well below the replacement level of 2.1%. China now has an ageing society, with 3% of its population aged over 60. The prospect ahead is one of a dwindling labour force, reduced income tax and increased social and medical costs, all of which could harm the Chinese economy in the future and put a burden upon the younger generations.

Negative evaluation. An ageing society.

Also worrying is a deepening imbalance in the sex ratio at birth, which rose from 116.9 boys per 100 girls in 2000 to 118.1 boys per 100 girls in 2010. That disparity suggests a spread in sex-selective abortion. China's gender imbalance is already severe, and its effects will only worsen with population momentum as those in the youngest birth cohorts become adults; the social implications of this are serious and long term. There have been several attempts to relax the policy, especially in rural areas, and couples who grew up without siblings are now allowed to have two children. However, 75% of the population support the one child policy, and urban living is considered incompatible with having more than one child. It may therefore be difficult to reverse this policy.

Negative evaluation. Gender imbalance.

The link between the one child policy and a successful economy is difficult to prove. It is possible that a demographic transition might have occurred in China without such severe intervention by government. This has been the case in many developing countries where more humane anti-natalist policies have been employed. For example, Kenya has developed economically despite a high rate of natural increase, and children there are considered to be an economic asset rather than a burden.

Negative evaluation. The weak link between declining fertility and growing wealth.

It conclusion, the one child policy was successful in arresting the rapid rate of population growth in China at the end of the twentieth century and in achieving a balance between resource supply and demand, but it was at a price and there are three negative outcomes: the abuse of human rights, the ageing population and the gender imbalance. All are serious problems with long-term negative outcomes.

A brief conclusion that gives an overview of the argument.

Key terms

Throughout your course and during your exams it is essential that when using concepts and terms you fully understand their definitions.

The following table outlines typical terminology you may come across, followed by definitions. It is important that you are clear about the precise meaning of such terms—many of which are commonly used in exams.

Concept	Definition
Contemporary	Refers to an event that has taken place within the student's lifetime.
Equilibrium	A state of balance among the components of a system.
Feedback	The return of part of the output from a system as input, so as to affect succeeding outputs. Negative feedback is feedback that tends to damp down, neutralize or counteract any deviation from a state of equilibrium, and promotes stability. In contrast, positive feedback is feedback that amplifies or increases change.
Geographic	All the demographic, environmental, social, cultural, economic and political and geopolitical factors that could influence or be influenced by the geography of an area.
Geopolitical	The combination of locational and political factors relating to or influencing a region or nation.
Intensity	This is the magnitude of an event such as a volcanic eruption or hurricane. It indicates its size and, very often, the amount of energy released. Volcanic intensity can be measured using the Volcanic Explosivity Index (VEI).
Interrelationship	How different features interact with one another. For example, the interaction between climate and vegetation in the generation of overland flow. If human activity removes some of the vegetation, then more run-off will occur as a result.
Less economically developed/more economically developed countries, rich/poor countries, North/South	These terms are not used in this syllabus but teachers may use them as they feel appropriate. An LEDC is a country with low to moderate industrialization and low to moderate average GNI per capita. In contrast, a MEDC is a highly industrialized country with high average GNI per capita.
Long-term	Factors, features or responses that occur much later, for example, redevelopment following an earthquake.
Model	A simplified description designed to show the structure or workings of an object, system or concept, such as the demographic transition model or Lee's migration model.
Pattern	This term covers the arrangement or changes of spatial elements (compare with "trend").

Concept	Definition
Recent	Reference to an event that has taken place since the year 2000.
Scale	Topics within the syllabus should be studied at various scales. • Local scale refers to areas of limited extent. • National scale refers to a country. • Regional scale refers to an area that encompasses several countries sharing some common element (economic, political, locational). It is larger than the country and smaller than the continent. • Global scale refers to the world as a whole.
Short-term	Factors or features that may happen quickly, for example, search and rescue following an earthquake.
Socio-economic	The combination of social factors (including demographic, cultural and political) and economic factors.
Spatial	Relating to area or place.
System	An assemblage of parts and the relationships between them, which together constitute an entity or whole. They tend to have inputs, stores, processes and outputs. A closed system is one in which energy, but not matter, is exchanged with its surroundings. In contrast, an open system is one in which both matter and energy are exchanged with its surroundings (for example, natural ecosystems).
Strategies	Reference to any management policies, initiatives and/or plans.
Temporal	Relating to time.
Trend	This term covers the changes over time (compare with "pattern").

3. Get to know your exam paper

This chapter will explain the structure of the exam and the details regarding each paper. It will also provide a few general guidelines that will help you on the different papers.

Paper 1 examines the core theme—patterns and change—and is taken by all students; paper 2 examines the optional themes (two optional themes are required for SL and three for HL); paper 3 examines the HL extension—global interactions (HL students only).

You will sit two exams (external assessments) for SL or three exams for HL. Written papers account for 75% of the marks for SL (with internal assessment making up the remaining 25%) and 80% of the marks for HL (with internal assessment making up the remaining 20%).

The contribution of each paper towards the final exam mark is as follows.

Standard Level

Paper	Time	Weighting	Marks
Paper 1 Section A: Answer **all** questions Section B: Answer **one** extended response question from a choice of three.	**1 hour and 30 minutes**	**40%**	**60** (45) (15)
Paper 2 **Two** optional theme questions. For each optional theme there is a choice of two questions.	**1 hour and 20 minutes**	**35%**	**40** (20 marks each)

Internal assessment

Internal assessment involves the completion of a 2,500-word report based on fieldwork.

The recommended class time to be devoted to fieldwork is 20 hours.

This component at SL counts for 25% of the total marks and is marked out of 30.

Higher Level

Paper	Time	Weighting	Marks
Paper 1 Section A: Answer **all** questions. Section B: Answer **one** extended response question from a choice of three.	**1 hour and 30 minutes**	**25%**	**60** (45) (15)
Paper 2 **Three** optional theme questions. For each optional theme there is a choice of two questions.	**2 hours**	**35%**	**60** (20 marks each)
Paper 3 Students answer **one** of three essay questions. Part (a) involves one short essay and one longer essay, which may or may not be connected to part (b). The two parts carry 10 marks and 15 marks respectively.	**1 hour**	**20%**	**25**

Internal assessment

Internal assessment involves the completion of a 2,500-word report based on fieldwork.

The recommended class time to be devoted to fieldwork is 20 hours.

This component at HL counts for 20% of the total marks and is marked out of 30.

Mark weighting

The table below shows the percentage of the total exam marks that are allocated to each component. This helps you to give sufficient attention, both in your revision and in the exam itself, to the papers and questions that have the greatest percentage weighting. For example, an HL student must ensure that they are well prepared for the essay they must write in paper 3, which is worth 20% of their total mark and, therefore, as much as their fieldwork report. In contrast, the essay question in paper 1 carries only 6.25% of the total marks and because there is no time limit for the separate completion of sections A and B, students may wish to allocate less time to the essay. However, if you are an SL student, you need to be aware of the importance of paper 1, which carries 40% of the total marks, with the essay carrying 10%.

Geography exam components—relative weighting

	% weighting of exam paper		**% weighting by section/question**	
Exam component	SL	HL	SL	HL
Paper 1 (A)	40	25	30	18.75
Paper 1 (B)			10	6.25
Paper 2	35	35	17.5	11.7
Paper 3	—	20	—	20
Internal assessment	25	20	—	—

Paper 1

Paper 1 is common to both SL and HL. It is divided into two sections—A and B—and its purpose is to test:

- knowledge and understanding (AO1) in sections A and B
- application and analysis (AO2) in sections A and B
- synthesis and evaluation (AO3) in section B
- the ability to use appropriate terminology and skill (where appropriate) in response to sections A and B, and to produce well-structured written material (AO4) in section B (see example below).

Section A

There are four compulsory short-answer questions covering the four topics within the core theme. Questions are subdivided into parts; the first part requires knowledge and understanding, while subsequent parts may require knowledge and understanding and application and analysis. Pay particular attention to the command words—these identify the depth of answer required. **A total of 45 marks are available in this section**. There could be as many as 12 questions in this section. Losing a couple of marks on each question can add up to a considerable loss—you should really focus upon the requirements of each question.

Section B

In section B, students answer one essay question from a choice of three. The question requires broad treatment and may require students to integrate topics across the core theme. This means that you will have to draw on your subject knowledge relating to at least two optional themes. The question may require knowledge and understanding, application and analysis, synthesis and evaluation. **A total of 15 marks are available in this section.**

It is recommended that you spend about 1 hour and 5 minutes on section A and about 25 minutes on section B—roughly 1.5 minutes for each mark available.

Paper 2 (two questions at SL, three questions at HL)

- This paper assesses assessment objectives 1–4.
- For each of the seven optional themes there is a choice of two questions, with each question being made up of three or four sub-questions.
- The first two or three parts add up to 10 marks, and the final part is a 10-mark extended response or essay-style question.
- The final part may not be related to the first two or three parts of the question.
- The command term indicates the depth of response that is required.
- The layout of the paper follows the order of the seven optional themes.
- The title of each theme is given in the exam paper.
- Students answer one question for each theme (two optional themes for SL, three optional themes for HL).
    - At least one of the two questions in each theme uses stimulus material (for example, maps, graphs, photographs, satellite images, diagrams and/or tables).
    - Stimulus material may be provided in the resource booklet.
    - **The maximum mark for each question is 20.**

Paper 3

- This paper assesses assessment objectives 1–4.
- The questions draw on information from the seven topics in the HL extension.
- Knowledge of the core theme is assumed.
- Students are required to answer one question from a choice of three; each question is split into two parts—part (a) is worth 10 marks and part (b) is worth 15 marks.
    - Part (a) tests knowledge and understanding, and application and analysis.
    - Part (b) tests synthesis and evaluation, and writing skills, in addition to knowledge and understanding, and application and analysis.
- You are expected to demonstrate synthesis and evaluation by making links and showing the relationships between the topics/sub-topics and by producing a structured response.
- You are advised to plan your response before writing. A plan can include:
    - key terms and definitions
    - main processes or factors
    - case studies integrated into the text, not as separate entities
    - a conclusion or view (depending on the wording of the question).
- The command term indicates the depth of treatment required.
- **The maximum mark available is 25 marks.**

4. Top tips

This chapter outlines some top tips to help you prepare for the exams and some extra tips to be aware of when actually taking the exams.

Know your syllabus

The *Geography guide* produced by the IB is a great place to familiarize yourself with the syllabus. If you do not have access to a copy, ask your teacher.

The syllabus is the foundation of the course—it lists the components of the course to be followed by students, teachers and examiners. Pages 4–10 explain the nature of the subject and the assessment objectives, which are essential reading for teachers and examiners.

The following pages are essential reading for you.

- Pages 12–14: Syllabus structure and important concepts used within the syllabus.
- Pages 15–18: Geographic skills: These are all the practical techniques required, many of which will appear in your exam papers.
- Pages 19–46: The syllabus content is divided into three parts:
  - Part 1: Core—patterns and change—for HL and SL
  - Part 2: Optional themes—two optional themes are required at SL; three optional themes are required at HL
  - Part 3: HL extension—global interactions—for HL only.

There is an introduction to each part that outlines the topic to be covered, the key concepts, the approach to learning and the general requirements.

Each sub-topic table is broken down into numbered sections and bullet points in the left column with further detail given in the middle column. Look carefully at the command terms used because they indicate the highest level of understanding required. For example, if you are asked to "Discuss the human causes and consequences of a specific river flood", you should be able to describe and explain them as well as discuss them. The "Teaching hours" column indicates the number of hours required to teach the sub-topic and shows its relative importance. Take particular note of case study requirements.

During the exam

- **Time mismanagement**: Every year, students lose marks through time mismanagement. The most common tendency is to spend too long on the first question at the expense of the others. Make sure that you are aware of the time allocation for each question and that you stick rigidly to it during the exam. Note that five minutes' reading time is allowed before the start.
- **Question choice**: In paper 1 and paper 3, all topics on the syllabus are compulsory, even though not all of them will be examined in one exam session. Again, your revision should be thorough and no sub-topic omitted. For example, in paper 1 you will need to cover all four areas of the core theme, but section A (short-answer questions) and section B (essays) will not necessarily cover all sub-topics in the syllabus. Familiarize yourself with the instructions at the top of your exam papers as these give you the instructions about timing and question choice.
- **Generalization (for example, "Africa")**: Geographers are keen on classification and putting phenomena into boxes. They also tend to generalize in their desire for pattern and order. However, generalization can sometimes be misleading, and statements such as "China is a developing country" fails in several respects. First, China is too large in physical size and population to classify in this way. Second, levels of development vary greatly between rural and urban areas. Third, the rate of economic change is likely to make any

classification obsolete within a few years. The solution is to choose a more reliable example, such as a smaller country at the bottom of the global economic development scale. Simplification can sometimes result in distortion of the truth. "Mumbai is a poor city" disguises the pockets of affluence and the existence of a very rich elite there. Expressions such as "e.g. Africa" used in exams reflect badly on the candidate. Using "etc" will earn no credit and suggests that you lack the knowledge or cannot be bothered.

- **Lack of evidence**: Essentially, geography is about the real world and, in longer responses, there must be plenty of factual support, examples and statistics. The geography of a fictional location is easily spotted these days and, if in doubt, the examiner will quickly check out the facts on the internet.
- **Outdated examples and case studies**: There are a few unique and well-documented case studies that are frequently used in exams. One very common example is that of the Lynmouth floods in Devon in 1952. However, there are more recent and better-documented examples of local flooding outside the UK. Outdated case study events are undesirable because human causes, consequences and responses have changed. Generally, the examples and case studies of geographic events occurring since 2000 are most relevant.
- **Lack of appropriate terminology**: Good use of geographical terminology shows understanding and avoids clumsy description. For example, a situation where "a population keeps on growing owing to a lot of young people who are likely to keep the birth rate high for some years into the future" could be described as "population momentum".
- **Inadequate reading of the question**: For example, if you were asked to comment on the relative advantages of tourism as a development strategy in a less-developed country and you discussed the issue in a more-developed country, you would inevitably lose marks. Ignoring question commands is common and is highly likely to lose marks.
- **Poor drawing skills**: The maps and diagrams produced by geography students are often small, badly drawn and lack information. These skills should be practised under exam conditions.
- **Missing the slant of the question**: Read the question three times; underline the command terms, other key words or slants of the question. To maximize your marks, you need to ensure that you have given the question its broadest interpretation. It is very easy to forget a category despite having plenty of knowledge about it. Causes, consequences, responses and factors may each be classified as shown below.
  - Positive or negative.
  - Long-term—geological timescale, century, decade, year. This depends upon the context.
  - Short-term—month, week, day.
  - Large scale/small scale—global, regional, international, national, local, household.
  - Physical—environmental, biological, geomorphological, geological, climatic, fluvial, coastal, tectonic or aesthetic.
  - Human—sociological, demographic, cultural, historic, economic or political.

Do

- Read the instructions on the cover of your exam paper to remind you of the exam regulations, such as the time allowed and the number of questions you should answer.
- Underline the command terms in the questions and focus on these as you work through each question.
- Write a brief plan for essays, to give your answer a logical structure. This should be written in pencil to allow for secondary thoughts or inspiration that comes to mind later.
- Observe the mark weighting of the parts of structured questions.
- Give sufficient attention to the parts of the question requiring evaluation, discussion or analysis. These carry a heavy mark weighting.
- Complete the correct number of questions.
- Make sure that all your answers are legible, correctly numbered and in numerical order.

Don't

- Pad your answer with irrelevant content just to make it look better. Examiners are impressed by quality, not quantity.
- Leave the examiner to draw conclusions if you cannot decide.
- Bend the question to fit your rehearsed answer.
- Spend too long on your best question at the expense of others.
- Invent case studies; these will be checked by examiners.
- Attempt to imply that part of an answer has been lost by exaggerating the number of pages completed on your front cover.
- Use lists or bullet points—these do not help detailed analysis.
- Make your own abbreviations, for example "U" for "urban" and "R" for "rural". You can use accepted abbreviations, but you must make sure you write each term out in full the first time you use it—for example, "infant mortality rate (IMR)" and "purchasing power parity (PPP)".

Look out for these dos and don'ts, as well as other common errors, in the student answers in this book.

5. Data response skills

Data response questions require you to interpret information or stimulus material such as tables, graphs, maps, oblique and aerial photographs and satellite images.

Paper 1 section A consists of short-answer questions, some of which will have stimulus material. Paper 2 consists of questions on the seven optional themes, each of which has two structured questions of which one or both will use stimulus material. The stimulus material will be found either within the question itself or in the resource booklet. You should become familiar with a range of stimulus materials and the techniques of data response through exercises and timed exam-style questions throughout the course. [Stimulus material is referred to throughout several of the following chapters.]

How to interpret thematic maps

Thematic maps are maps, at any scale, which show one or more geographical aspects, such as temperature, desertifiation and life expectancy.

You should be able to:

- recognize and describe the broad general patterns, as well as specific detail and anomalies
- analyse patterns and distributions, and relate these to named regions/countries
- use the key effectively in order to quantify your answer.

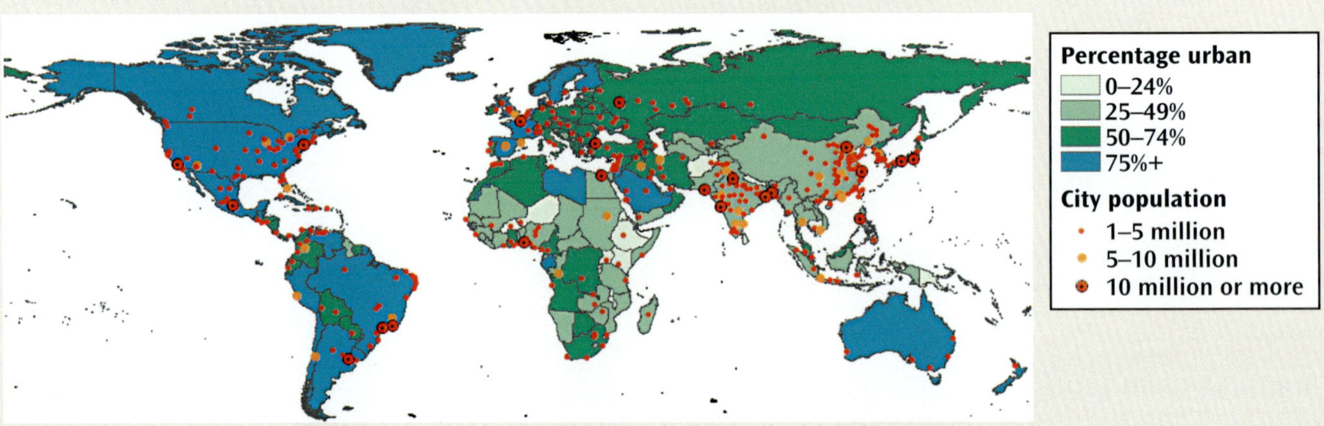

Source: Department of Economic and Social Affairs, Population Division. 2010. *World Urbanization Prospects, the 2009 Revision*. New York, NY, USA. United Nations.

Examples

> Describe the **pattern** of urbanization shown on the map.

"Pattern" means the spatial variation in levels of urbanization.

First, you should describe the overall global pattern by briefly identifying the link between development and the level of urbanization, that is, that high-income countries have high levels, low-income countries have low levels and middle-income countries have intermediate, but more varied, levels of urbanization. Next, provide evidence from the map giving region or country names and statistics to support these statements. For example, levels are highest (more than 75%) in North America, north-west Europe and Australia, and lowest in the poorest regions of Sub-Saharan Africa and South-East Asia (less than 25%). Levels of urbanization in middle-income countries, such as those in South-East Asia and South America range between 25% and 50% and it is more difficult to generalize. For example, many South American countries have levels above 75%. This might be considered an exception to the rule or an anomaly.

This is a long answer, and the amount of detail you give in this answer will depend upon the marks and time available.

> Explain the **distribution** of cities of different population sizes shown on the map.

"Distribution" means the spread of cities over an area.

Although this question does not ask you to describe the location of cities of different sizes, it is essential that you do this, if only very briefly. For example, the largest cities with a population of over 10 million are found in every continent, but especially in the mid and lower latitudes, and many are coastal, as in South America. Smaller cities are found in similar locations, but the smallest with a population under 5 million are widely distributed, as in China.

In the more-developed world, such as north-west Europe and North America, the long-standing urbanization process has resulted in a developed hierarchy with a range of different-sized cities. In many middle- and lower-income countries, large cities have developed in coastal locations as a result of their historical trading connections. These connections have encouraged recent and rapid urbanization through both internal and external immigration, for example, Cairo, Mumbai and Manila.

How to interpret topographical maps

Topographical maps have contour lines showing altitude and relief features. In order to use these maps effectively, you need to understand and use the techniques given below.

These maps may appear in any optional theme within paper 2, although they are most likely to be seen in the following optional themes.

- Option A: Freshwater—issues and conflicts
- Option B: Oceans and their coastal margins
- Option C: Extreme environments
- Option E: Leisure, sport and tourism
- Option G: Urban environments

You should be able to:

- calculate distances on a map using the representative fraction, such as 1:50,000.
- estimate areas
- estimate directions using compass readings
- use grid references.

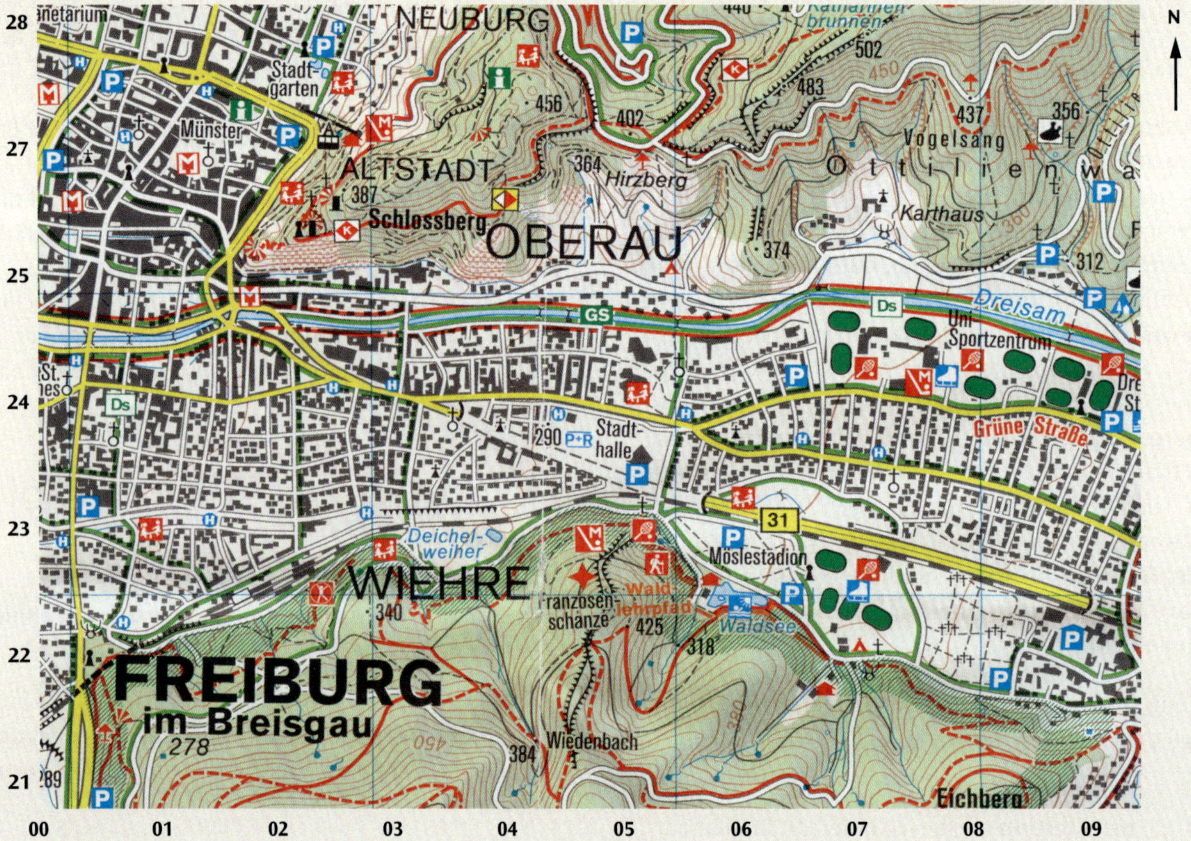

Scale 1:25 000

0 — 1 km

- Tennis court, indoor tennis court
- Golf, miniature golf
- Hang-/paragliding, open air aviary
- Forest mile/fitness course
- Climbing wall, ice climbing wall
- Playground, bowling
- Stable, game park
- Spa, protohistoric archeological dig
- Shooting range, holiday village
- Summer toboggan run, botanical garden
- Model flight area, sunbathing lawn
- Information
- Alpine hut/inn, catered in summer and winter
- Alpine hut/inn, catered in summer
- Hotel, inn, restaurant
- Restaurant, snack-bar
- New wine pub
- Skate park, leisure facilities
- Bike repair shop, ice-skating rink
- Indoor swimming pool, outdoor swimming pool/lake swimming
- Spa, 'Kneipp' facility
- Sailing, surfing
- Boat rental, rafting
- Fishing, cross-country skiing
- Toboggan run, curling
- Barbecue area, bicycle hire
- Sports ground, ski jump
- Open air stage, experience park
- Wayside cross, tree with an icon
- Castle/palace, ruin
- Convent, monastery
- Lookout tower
- Monument
- Forester's house
- 885 Elevation mark

Source: Geobasisdaten©Landesamt für Geoinformation und Landentwicklung Baden-Württemberg, (www.lgl-bw.de); AZ: 2851.2 – D/7848

Scale

The scale of this map is 1:25,000.

This means that 1 cm on this map represents 25,000 cm in reality or that 4 cm on the map represents 1 km.

Grid references

Most maps are dissected by gridlines. Each grid on this map is 1 km square, but because they are feint, the axes of the map have been numbered to help you locate features accurately.

Select the gridline to the west and the gridline to the south of a map feature. Give a three-digit reading by estimation along each side of this grid square.

For example, the information centre in Freiburg is in grid square 01 27. Its six-digit reference is 017 273.

Direction

The ice-skating rink at Möslestadion is 2.6 km south-east of the information centre in Freiburg.

The map extract shows the centre and south-west part of Freiburg, an historic city with a population of 222,000 and situated in south-west Germany.

Referring to the map, explain the pattern of leisure and recreational land use in central and south-east Freiburg.

There is a concentration of tourist facilities within the central business district (CBD), which has an historic core and is located between eastings 00 and 03 and northings 25 and 28 and is recognized by the high building density there. Tourist attractions include many museums, and there is a tourist information centre at 017 273. These facilities have a high threshold population and a national or international range. Consequently, they require a central location for easy access by tourists and they can afford the high land values of the CBD. Other facilities, such as a skating rink and sports grounds, require cheaper land and more space for construction. They are found 2–3 km east of the CBD where they are likely to serve the residential population rather than tourists. Further out, beyond the urban fringe, land uses relating to rural recreation and tourism can be found. These include walking trails on the hills north-east and south-east of the city at Oberau and Wiehre respectively.

How to interpret graphs

Scatter graphs

Geographical data is often presented in a graph, which shows relationships. A scatter graph, such as the one opposite, is commonly used, and you need to be familiar with this type of graph. A further complication here is the use of a logarithmic scale. This allows a greater range of values to be accommodated, but reading actual values off the graph can be difficult. If you are quoting data from the graph, try to pick examples that are close to a gridline and, therefore, more easily estimated.

GDP per capita and school life expectancy, 2006

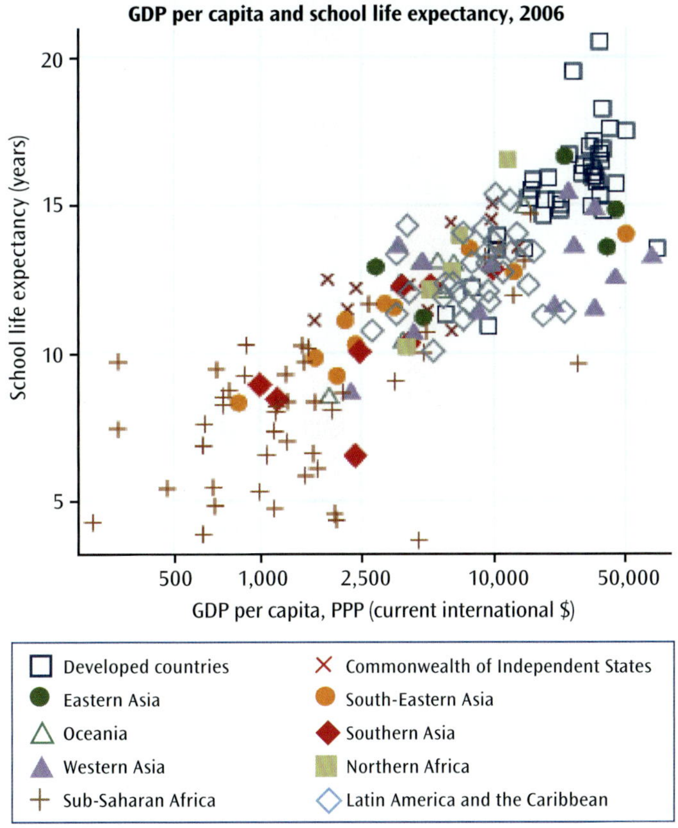

Data sources: UNESCO Institute for Statistics, World Bank, UN Population Division
Friedrich Huebler, "National wealth and years of education", International Education Statistics, August 2008, http://huebler.blogspot.com/2008/08/wealth.html
Trevor Cole/www.coleimages.org

Note: School life expectancy is the number of years a child is expected to have in basic education.

You should be able to:

- describe the relationship between school life expectancy and GDP.
  - There is a positive relationship (or correlation) between school life expectancy and GDP per capita.
  - Poor regions have low school life expectancy—under 10 years—and low GDP—under US$2,500.
  - Rich countries have high school life expectancy—over 15 years—and high GDP—over US$25,000.
  - Mention anomalies.

Compound bar graphs

Compound bar graphs are used to show the relative expenditure on different types of disaster.

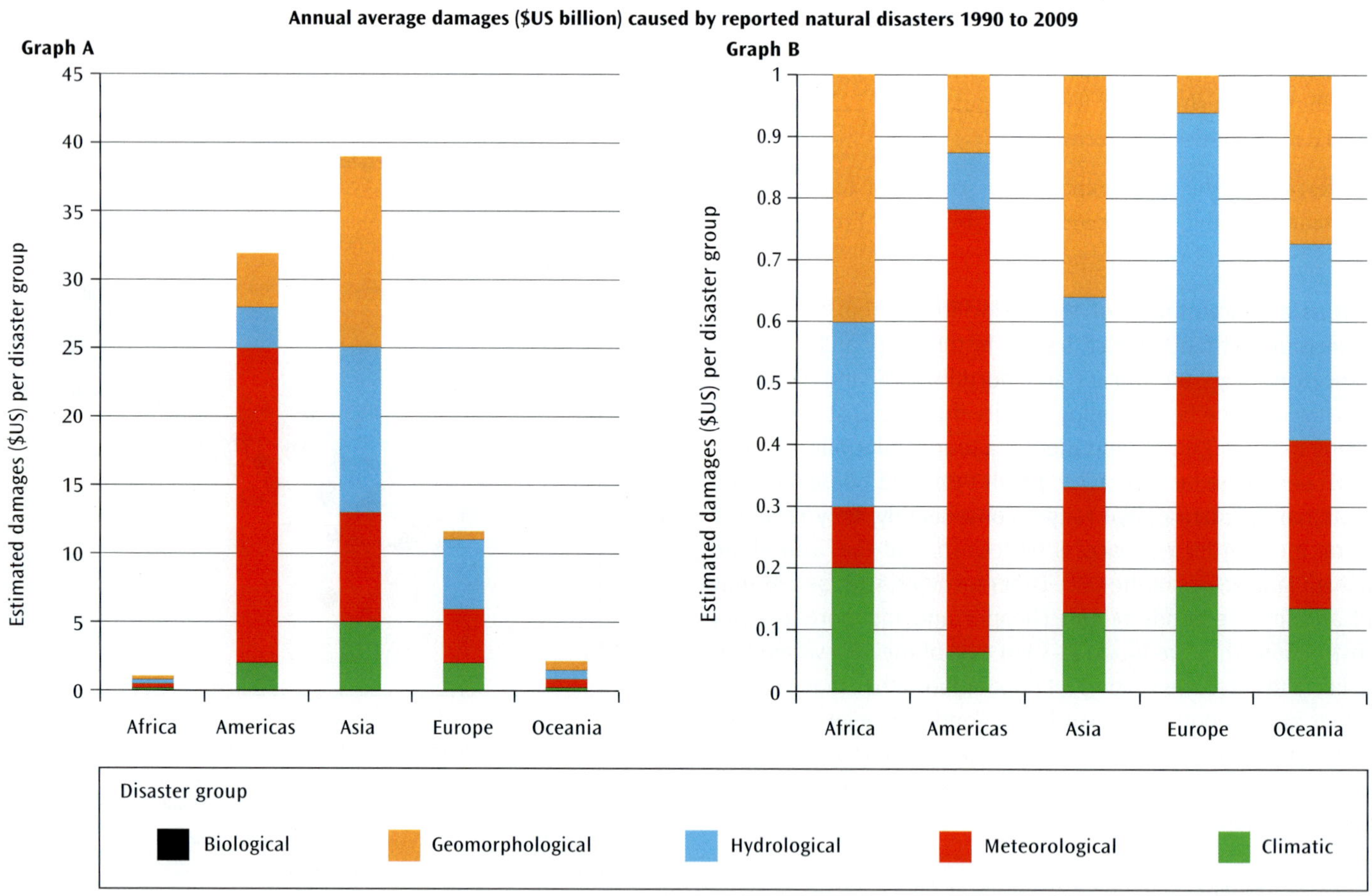

Source: EM-DAT: The OFDA/CRED International Disaster Database—www.emdat.be—Université Catholique de Louvain—Brussels—Belgium.

Graph A—total spending between regions can be easily compared but some of the smaller values for each disaster are difficult to see.

Graph B—the relative importance of expenditure by disaster can be compared between regions, but not the total expenditure.

Multiple line graphs

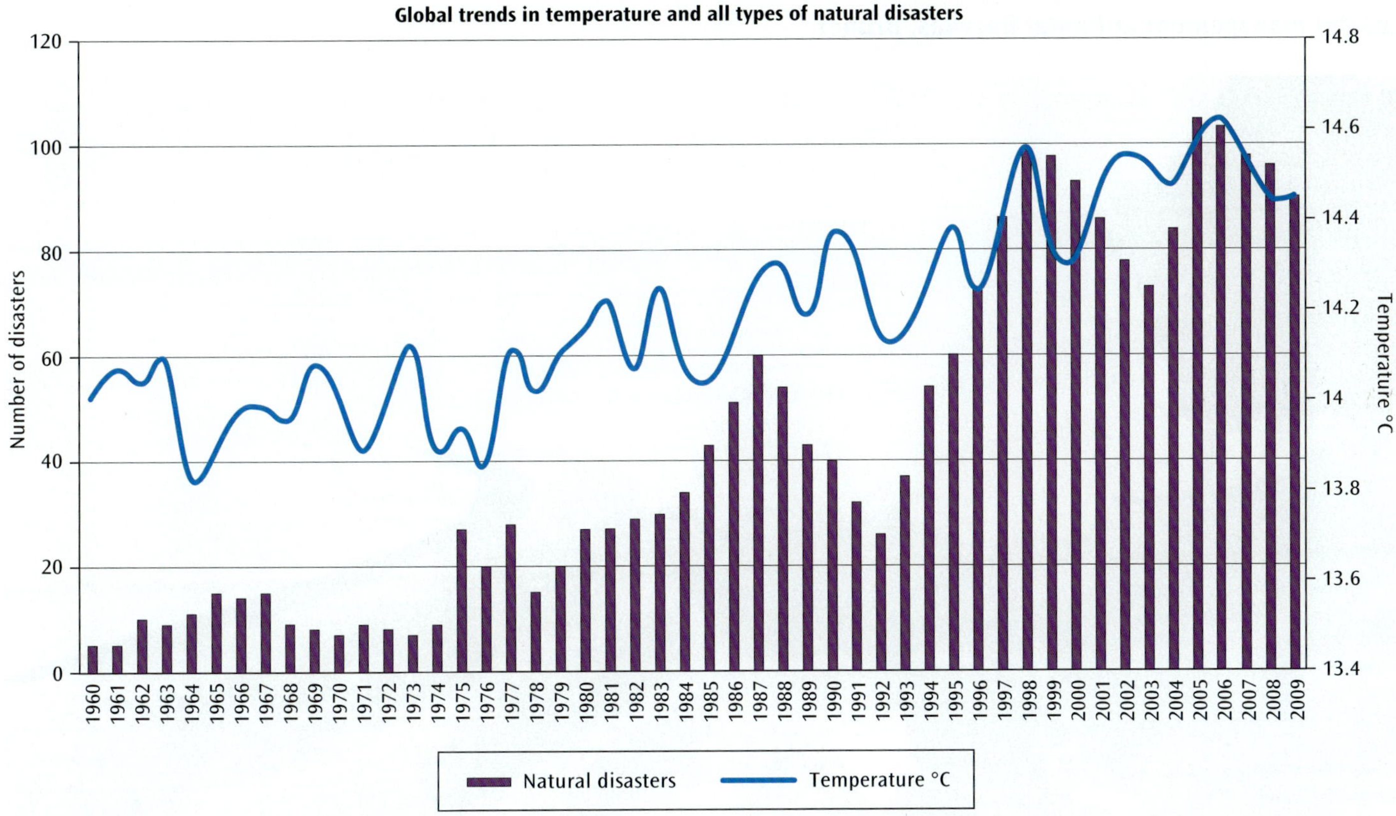

Sources: Adapted from The Goddard Institute for Space Studies Surface Temperature Analysis (http://data.giss.nasagov/gstemp/)
The OFDA/CRED International Disaster Database http://www.emdat.be/natural-disasters-trends

You should be able to:

- describe and explain the trends in world temperature and natural disasters between 1960 and 2009
- suggest possible reasons for the trends shown on the graph.

Both variables show an upward trend between 1960 and 2009, but with an increasing rate in the case of disasters, which is particularly marked in the years since 1990. Over 49 years, temperature has increased by 0.5°C and the number of natural disasters has risen from three in 1962 to a peak of 115 in 2005. Note that the axis representing temperature has been adjusted to conform to that of natural disasters and to give an impression that they are associated.

The upward trend for both global temperatures and disasters suggests that they might have some link or common cause. It is possible that the rising temperature resulting from global climatic change may result in an increasing frequency of climatic hazards such as droughts, floods, bushfires, urban heatwaves, storms, tropical cyclones, storm surges, avalanches and mudslides. However, data here also represents other hazards such as earthquakes and volcanoes which are not linked to climate.

The increasing incidence of such disasters may be related more to increasing population, a higher level of vulnerability and better reporting than to a higher frequency of natural hazards.

How to interpret a photograph

Coastal management at Pantai Berakas, Brunei

Source: Garrett Nagle

Photographs are another form of stimulus material but, in the exam, you will not be able to mark the photo itself and you will have to give the examiner clear directions to features in the photo. There are two approaches to this problem.

- Locate the features you are describing by using expressions such as "foreground", "background", "centre", "left" and "right".
- Draw a labelled sketch of the photograph.

Example

> Referring to the photograph, describe and explain the techniques used to control erosion on this stretch of coastline.

The purpose of management of this coastline is to control the process of coastal erosion.

The gabion cages at the cliff base (centre foreground) present a rough, resistant surface to oncoming waves to absorb wave energy.

The soft clay cliffs have been regraded and covered with concrete to maintain their stability (centre right).

The possibility of mass movement resulting from added weight on the cliff face has been reduced by drainage. The removal of water has been achieved by drainage of the cliff foot, construction of a drain between the gabion cages and the cliff face, and by dredging the stream channel.

Natural management (soft engineering) has been used to maintain forests, which absorb wave energy and offer some protection to the coastline.

How to draw a sketch

The table below outlines some key guidelines on drawing annotated sketches.

Do

- Draw a large diagram filling at least half a page.
- Place your annotations around the diagram and point them to specific places.
- Write concisely.
- Use pencil and colours.

Don't

- Separate your diagram from your annotations.
- Separate your diagram and annotations from the rest of your exam paper.
- Attempt this question if you are slow at drawing.
- Add a diagram at the end of an answer—it makes the answer appear unfinished.

Annotations and labels

You may be required to annotate or label your diagram or map. An annotation is a short description or explanation; a label is usually just one or two words to identify the feature. Both annotations and labels should be written on the diagram or map.

If space is a problem, you may use a key with arrows pointing to the correct location on the diagram or map. You should not write these labels or annotations separately below the map or on the next page of your answer.

Example

Using only an annotated diagram, explain the formation of floodplain features.

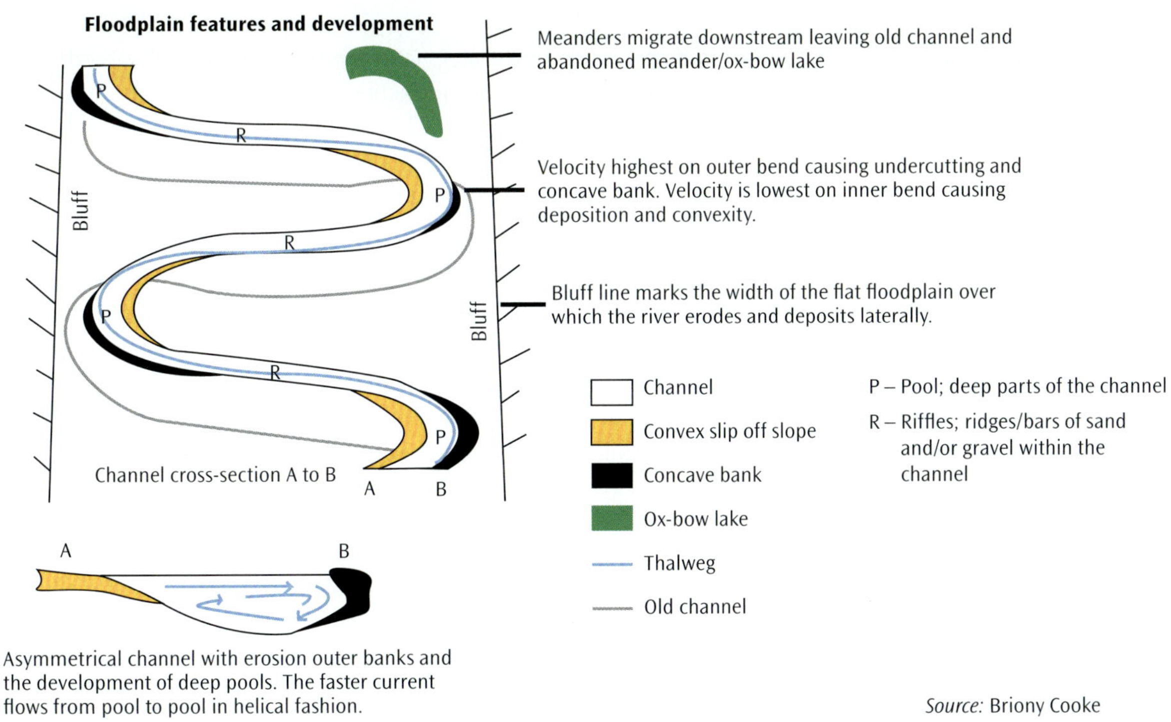

Source: Briony Cooke

6. The core theme

The core theme—patterns and change—provides an overview of many contemporary global issues. It can be applied on a range of scales, from national to regional to global.

Inequities are examined through the study of the United Nations' Millennium Development Goals (MDGs), and progress towards achieving these goals is investigated. The approach to teaching these global issues should be optimistic and positive. It is recommended that as much of the content as possible is taught through appropriate examples and case studies, preferably chosen from a limited number of countries (between three and five) of contrasting levels of development, for example, rich and poor countries, emerging economies.

Section A

Populations in transition

You should be able to:

- explain population trends and patterns in births (crude birth rate), natural increase in mortality (crude death rate, infant and child mortality rates), fertility and life expectancy in contrasting regions of the world
- analyse population pyramids
- explain population momentum and its impact on population projections.

You should know how each of these indicators is determined, be able to describe and give reasons for the trends (changes over time) that have occurred in world regions that contrast in terms of development. For example, you should know statistics for all indicators for contrasting regions and should be able to give at least one example of a country within each region. You should be able to recognize population patterns and explain them. A common error is to assume that high rates of natural increase are explained by high birth rates alone. Remember that it is the relationship between birth rate and death rate that determines natural increase, while the relationship between birth rate, death rate, immigration and emigration determines population change.

You should be able to draw two contrasting pyramids and show their development over time.

Features such as the influx of immigrants, the impact of birth-control policies and gender imbalances could all be explained through annotations on a population pyramid. (Refer to chapter 5, "Data response skills", for advice on drawing an annotated diagram.)

You should be able to:

- explain dependency and ageing ratios
- examine the impacts of youthful and ageing populations
- evaluate examples of a pro-natalist policy and an anti-natalist policy.

Dependency ratios are found by calculating the proportion of young and/or elderly people relative to the working population (see the "Key definitions" on page 26).

You should be able to explain both positive and negative socio-economic impacts for both ageing populations and youthful populations.

$$\text{Dependency ratio} = \frac{\text{population aged 0–19 years and population aged 65 years and over}}{\text{Population aged 20–64 years}} \times 100\%$$

You should be able to:

- discuss the causes of migrations, both forced and voluntary
- evaluate internal (national) and international migrations in terms of their geographic (socio-economic, political and environmental) impacts at their origins and destinations.

Bear in mind that you need to have selected case studies that fulfill these requirements and that give the topic its broadest interpretation. This means looking at migrations with different motives, on different scales and with a range of impacts, which may be positive or negative and long or short term. Your case studies should not be confined to the rich world because both internal and international movements have profound socio-economic effects in the poor world.

You should be able to:

- examine gender inequalities in culture, status, education, birth ratios, health, employment, empowerment, life expectancy, family size, migration, legal rights and land tenure.

The implication here is that the status of women is inferior, although it is not specified. You should explain the origins of inequality, its consequences, and any attempt to achieve a balance of opportunities. In order to cover all aspects of gender inequality listed here, you should select two countries with marked cultural contrasts. For example, one selected from the Middle East, where gender inequality is still an issue, and one selected from northern Europe. You should know some appropriate statistics, such as the gender gap index.

Key definitions

- **Child mortality rate**: The total number of deaths of children aged 1 to 5 years per thousand children aged 1 to 5 per year.
- **Crude birth rate**: The average number of births per thousand of population per year.
- **Dependency ratio**: The proportion of the population aged under 15 and over 65 (the economically dependent) relative to the proportion of the population aged 16 to 65 (the economically active).
- **Fertility rate**: The average number of children per woman.
- **Infant mortality rate**: The total number of deaths of children less than 1 year old per thousand live births per year.
- **Life expectancy**: The average number of years that a person can be expected to live, given that demographic factors remain unchanged.
- **Population momentum**: The tendency for a population to grow despite a fall in fertility levels because of a relatively high concentration of people of childbearing or pre-childbearing years.

The graph shows how female empowerment in politics has changed in different regions since 1997.

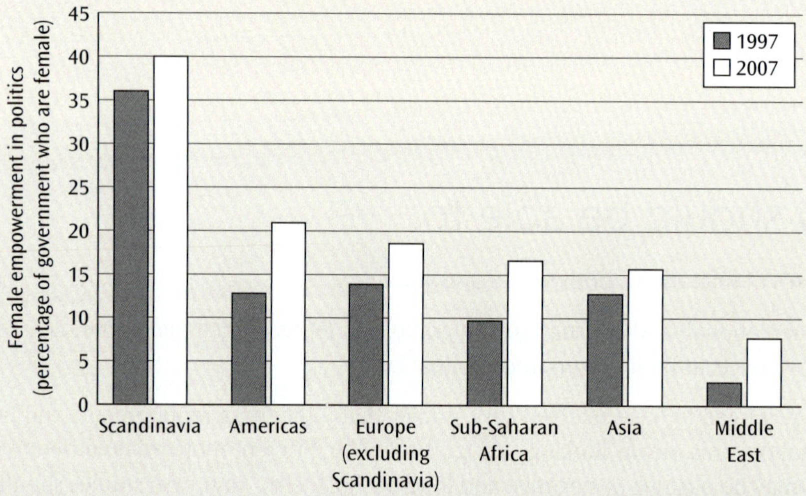

Source: adapted from UNESCO

(a) Referring to the graph, describe how female empowerment in politics varied between regions in 1997. *[3 marks]*

(b) Suggest reasons why female empowerment in politics has risen since 1997. *[4 marks]*

(c) Analyse the pro-natalist policy of a country you have studied. *[5 marks]*

[Taken from paper 1, May 2011]

How do I approach the question?

This question is divided into three parts, two of which relate to the graph.

(a) You should make sure that you use information from the graph and use quantification to give your answer some detail. For example, Scandinavia, as a region, has the highest level of female empowerment, with approximately 35% of those involved in government being women. The Middle East has the lowest, with only about 3% involved in politics. A list or an answer without quantification will score a maximum of two out of a possible three marks.

(b) You should either suggest four valid reasons or give two reasons in depth. Reasons include cultural, educational, social (for example, family size) and legal ones.

(c) You are asked to analyse—this means that you must break down the topic to make clear the essential elements of, in this case, a pro-natalist policy. Your answer should include a named country and policy, and a description of the policy. The final two marks of this question are reserved for an evaluation of the policy.

What are the key areas of the syllabus?

- The core: Populations in transition
- Sub-topics: Gender and change; Responses to high and low fertility

This answer achieved 11/12

The student has successfully described how female empowerment varies and has used supporting evidence from the graph.

(a) Female empowerment varies between the regions; in 1997 Scandinavia was the region with the highest numbers of women empowered in politics with a figure of about 36%. In the rest of Europe this figure was considerably lower 14% and a similar figure was in America and Asia 13%, slightly lower is the percentage of women empowered in politics in Sub-Saharan Africa while in the Middle East the percentage (3%) is extremely low. [3]

The student gives four valid reasons. The student could have scored four marks by writing about two in detail or four in less detail.

(b) Female empowerment in politics has risen since 1997 in all over the world because as women obtained the right of voting in different areas of the world they sustained the necessity of a female representative. As measures to reduce gender inequalities women obtained equal instruction opportunities and therefore the opportunity to perform any kind of job especially in developed countries women tend to concentrate more on their careers rather than producing their children as in the past. Medical technology improved women do not die giving birth therefore percentage of women working in politics will increase. [4]

This answer offers some evaluation but is not detailed enough. A more sophisticated answer might have stated "In Japan, companies have been encouraged to allow their employees to have more time off to procreate; in addition, shops offer discounts for larger families, and government child allowances are being introduced in an attempt to raise birth rates."

(c) In Russia in order to fight the demographic depression a sex day has been established the 12th of June every year. If one couple has a child born exactly nine months from the 12th of June his family will get significant economic benefits and other material compensation. Even if this policy might encourage people to get involved into procreation process during a day of the year the fact that many families will try to get funded would pull some couples out of the competition adding other dates like the 12th June could have the effect of increasing the Russian youthful population even because one day a year does not seem to be enough in order to achieve this target. [4]

Disparities in wealth and development

You should be able to:

- define indices of infant mortality, education, nutrition, income, marginalization and Human Development Index (HDI)
- explain the value of the indices in measuring disparities across the globe.

You should be able to define each indicator and provide statistical examples of extreme values in all cases.

You should be able to:

- explain disparities and inequities that occur within countries resulting from ethnicity, residence, parental education, income, employment (formal and informal) and land ownership.

When explaining disparities and inequalities, the scale used is national, which means that you are looking at social variations within one country and you should be able to name specific locations in your answer. For example, South-East China has a much higher level of development than in western provinces of China.

You should be able to:

- identify and explain the changing patterns and trends of regional and global disparities of life expectancy, education and income
- examine the progress made in meeting the Millennium Development Goals (MDGs) in poverty reduction, education and health.

You should be able to interpret spatial patterns on a world map and interpret trends over time on a graph. (Follow the advice on map interpretation given in chapter 5, "Data response skills".)

Choose two countries with contrasting rates of progress towards the Millennium Development Goals. Explain the disparities between these two countries; the barriers to progress and incentives.

You should be able to:

- discuss the different ways in which disparities can be reduced with an emphasis on trade and market access, debt relief, aid and remittances
- evaluate the effectiveness of strategies designed to reduce disparities.

This section requires an overview of global inequalities and how they might be reduced. A case study of one particular region or country in Africa, central Asia or South America would exemplify this well. The concept of internal and external forces determining global disparities in wealth is fundamental to this question.

Key definitions

- **Aid**: The transfer of money, food, goods, technology and education from one country, institution (World Bank, IMF) or NGO/charity (Oxfam) to a country in need.
- **Remittances**: Transfers of money/goods by foreign workers to their home countries.

The map shows the Gross National Income (GNI) per person for different countries in 2008.

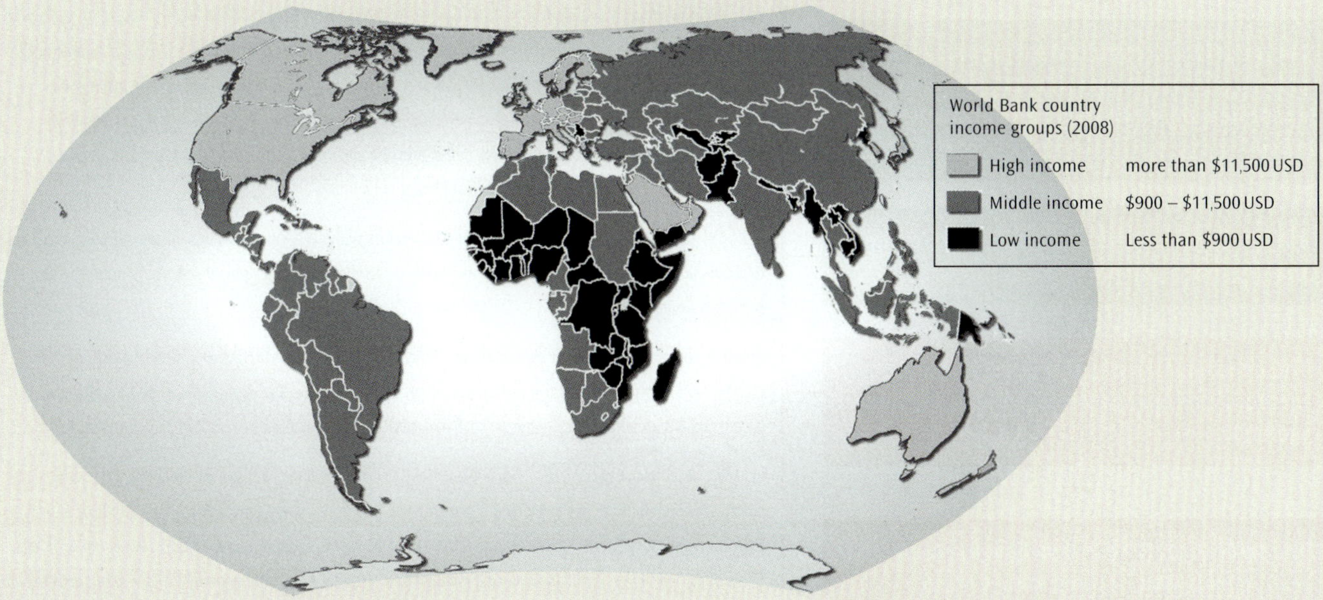

Source: Reprinted with permission from UNEP/GRID-Arendal, http://maps.grida.no/go/graphic/world-bank-country-income-groups

(a) Describe the global pattern of GNI per person shown on the map. *[3 marks]*

(b) Suggest **two** reasons why GNI per person is not a reliable way of measuring global disparities. *[2+2 marks]*

(c) Explain how debt relief may help to reduce global disparities. *[5 marks]*

[Taken from paper 1, May 2011]

How do I approach the question?

This question is in three parts—part (a) involves using a map as a stimulus, part (b) asks for a critique of the method of measuring GNI per person, and part (c), which is more difficult, asks for an explanation of how debt relief can help reduce global disparities.

(a) You should refer to the information shown on the map. A useful method is to identify the areas with the highest income (and state what is meant by a high income—use the scale) and the areas with the lowest income (again, with data). Obvious anomalies could also be identified, such as Saudi Arabia. An answer without quantification and/or anomalies can achieve a maximum of two marks.

(b) Two reasons should be identified and explained. For example, GNI per person is not a composite indicator; it does not show spatial variations within a country and does not take account of the environmental damage produced in generating wealth.

(c) Answers should begin with a definition of debt relief and then explain how debt relief could benefit countries. Answers that use examples of countries or regions that have experienced debt relief should be included. Alternatively, you could argue that debt relief does not reduce global disparities. Again, you should provide examples to back up your argument.

What are the key areas of the syllabus?

- The core: Disparities in wealth and development
- Sub-topics: Measurements of regional and global disparities; Reducing disparities

This answer achieved 8/12

The student successfully describes the global pattern and uses map evidence. Although quantification is not used here, anomalies are identified.

(a) This map shows how countries with a high income are concentrated within the first world, in particular North America, Europe and some country in the Arabic Peninsula. Other exceptions of high income are in Oceania, Australia and in the Far East, Japan and South Korea. Countries with a middle income are located in South and Central America, South and North Africa Asian and Eastern Europe. Countries with a low income are located in Africa in the Sahel Region and in central Africa there are some exception in Asia like in IndoChina, Central Asia and Pakistan. [3]

The student identifies one reason and implies that it is not a composite measure.

(b) GNI per person only measures the economic factors without taking into account any other figures linked to standard of livings, life expectancy or any kind of social disparities or gender inequality present within a certain country.

The question asks why GNI is not a reliable method. This answer states that it is a reliable method.

Another reason why GNI per person alone is a reliable way of measuring global disparities is that the income per person in a country is not an indication of the resources the population of that country can access and the quality of that resource furthermore is not an indication of freedom or political equality. [2]

The student could have benefited by saying what debt relief is, and by using some evidence.

(c) Debt relief may reduce global disparities as will reduce economic pressure on the third and fourth world countries which have to pay extremely high debts and therefore cannot concentrate their finances in providing their own people better standards of living. Debt relief reducing economic pressure on developing countries could allow for these countries to place the bases for a stronger economy and could allow many countries to extract by themselves important economic resources they couldn't extract before the new resources to trade will make those countries stronger on the global market. [3]

Patterns in environmental quality and sustainability

You should be able to:

- describe the functioning of the atmospheric system in terms of the energy balance between solar and long-wave radiation
- explain the changes in this balance due to its external forcings (changes in solar radiation, changes in the albedo of the atmosphere and changes in the long-wave radiation returned to space)
- discuss the causes and environmental consequences of global climate change.

Your answer can be enhanced by a well-labelled diagram showing radiation balances in the atmosphere.

You should be prepared to discuss the implications of changes induced through human activities.

Remember that this section focuses upon global climatic change rather than just global warming.

You should be able to:

- explain the causes of soil degradation
- discuss the environmental and socio-economic consequences of this process, together with the management strategies.

You should understand the general processes leading to soil degradation together with the interplay of human and physical factors.

The African Sahel provides several examples of countries where difficult physical circumstances are exacerbated by exploitative human activities resulting in soil degradation.

You should be able to:

- identify the ways in which water is utilized at the regional scale
- examine the environmental and human factors affecting patterns and trends in physical water scarcity and economic water scarcity
- examine the factors affecting access to safe drinking water.

You will need to show an appreciation of the relationship between water utilization and wealth.

You should be able to define both physical water scarcity and economic water scarcity, and provide an example of each. You should also be able to recognize the regional pattern of water utilization (this means global regions, such as Asia-Pacific or Latin America). You should also be able to explain the trends in water utilization over time.

You should be able to:

- explain the concept and importance of biodiversity in tropical rainforests
- examine the causes and consequences of reduced biodiversity in this biome.

You should know the location of tropical rainforests within the equatorial zone, and its associated climatic characteristics. You should also be able to describe the layers of the rainforest and its various ecological niches.

Loss of biodiversity should be explained through at least two examples, citing the value of biotic materials and the long-term consequences of their destruction.

You should be able to:

- define the concept of environmental sustainability
- evaluate a management strategy at a local or national scale designed to achieve environmental sustainability.

There are a number of possible case studies you could use from your local area, the most likely being strategies of waste recycling or energy and water conservation.

Key definitions

- **Biodiversity**: This term means biological diversity. It is the variety of all forms of life on Earth—plants, animals and microorganisms.
- **Economic water scarcity**: Where water is available locally but not accessible for human, institutional or financial capital reasons.
- **Enhanced greenhouse effect**: A result of human activity (burning fossil fuels), which increases the amount of atmospheric greenhouse gases and causes heating of the atmosphere.
- **Global climate change**: The changes in global patterns of rainfall and temperature, sea level, habitats and the incidences of droughts, floods and storms, resulting from changes in the Earth's atmosphere, believed to be mainly caused by the enhanced greenhouse effect.
- **Physical water scarcity**: Where water resource development is approaching or has exceeded unsustainable levels; it relates water availability to water demand and implies that arid areas are not necessarily water scarce.
- **Soil degradation**: A severe reduction in the quality of soils. The term includes soil erosion, salinization and soil exhaustion (loss of fertility).

The graph shows global temperature changes since 1850.

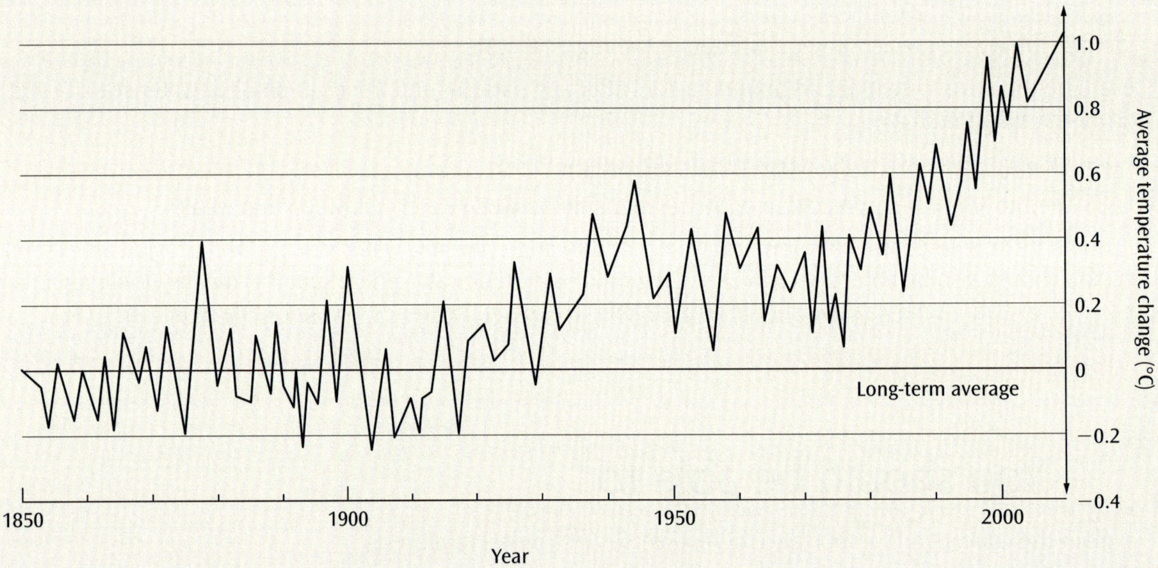

Source: IPCC—data based on the graph in *The Economist*, 9 September 2006

(a) State **one** natural external forcing which could have influenced the changes in temperature shown on the graph. *[1 marks]*

(b) Referring to the graph, describe how temperature has changed since 1850. *[3 marks]*

(c) Explain **three** possible environmental consequences of global climate change since 1950. *[3×2 marks]*

[Taken from paper 1, May 2011]

How do I approach the question?

The stimulus for this question is a graph of temperature changes since 1850. Be careful to read the scales—the vertical axis shows average temperature **change** from 1850 not average annual temperature.

Part (a) asks you to **state** one natural external forcing—for this, you only have to name a natural external forcing, for example, sunspot activity, changes in solar radiation, changes in albedo.

In part (b) you have to refer to the graph. Look for trends in temperature change, for example, is there a period when temperature: increases rapidly; is steady; falls; increases slowly? When is it below the long-term average? When is it above the long-term average? One mark is reserved for quantification or identifying the fluctuating nature of temperature change.

Part (c) requires you to explain three possible environmental consequences. Note that the focus is on environmental consequences of global climate change, not economic, social or political consequences. Two marks are available for each of the three consequences. Consequences include a rise in sea levels, expansion of some deserts, increased frequency and intensity of tropical storms, melting of the Arctic ice, retreating glaciers.

What are the key areas of the syllabus?

- The core: Patterns in environmental quality and sustainability
- Sub-topic: Atmosphere and change

This answer achieved 7/10

The answer must be external to the Earth's atmosphere, for example, changes in solar radiation (sunspot activity), changes in orbital patterns.

(a) Changes in temperature from 1850 to 2000 might have been influenced by the CO_2 emissions increase causing the greenhouse effect. [0]

Three valid points with quantification.

(b) According to the graph temperature change since 1850 has been floating around the same levels up to the 1930's. From the 1930's temperature begins gradually to increase to 1950, temperature decreases slightly from the 1950s to the 1980s however from the 1980's starts to increase rapidly up to its highest year in 2000. From the graph its estimated that the temperature will keep increasing. [3]

This is a cause not a consequence.

(c) One of the consequences could be the greenhouse effect CO_2 gasses form a barrier in the earth's atmosphere which does not allow the heat out so the planet is gradually warming up like a greenhouse

Two valid consequences with added detail and/or explanation.

The North Pole Ice's are melting always more rapidly during the summer and are not freezing as in the past during winter. This is causing ocean lakes to rise and most of the land where we live is under threat.

Seasons are changing as the temperature continues to rise and animals and plants have their natural cycle altered. This affects the harvests and every ecosystem in which we live. [4]

Patterns in resource consumption

You should be able to:

- evaluate the ecological footprint as a measure of the relationship between population size and resource consumption
- identify international variations in its size
- discuss the two opposing views (neo-Malthusian and anti-Malthusian) of the relationship between population size and resource consumption.

You should be able to define the ecological footprint and assess its value as a measure of the relationship between population and resources. International variations in size are best identified through reference to contrasting examples. **It is important to remember that the ecological footprint is not the same as the carbon footprint, a term that is commonly used in the media.**

You should know when Malthus wrote, and the basis of his theory. An annotated graph showing the relationship between population and food would illustrate this effectively. You should be able to explain the theories of the pessimistic neo-Malthusians, such as the Club of Rome and Paul Ehrlich, and the ever optimistic anti-Malthusians, such as Julian Simon. Relevant concepts include those of carrying capacity and sustainable resource management.

You should be able to:

- examine the global patterns and trends in the production and consumption of oil
- examine the geopolitical and environmental impacts of these changes in patterns and trends
- examine the changing importance of other energy resources.

Be prepared to identify, on a map, countries with high levels of oil production and be able to explain the physical circumstances leading to its availability. You should be able to relate the patterns of global production to consumption.

In addition to the spatial pattern, you should be able to explain the trends in oil production and consumption over time and changes in the energy mix for many countries.

You should be able to:

- discuss the reduction of resource consumption by conservation, waste reduction, recycling and substitution
- evaluate a strategy at a local or national scale aimed at reducing the consumption of one resource.

You should know the general principles of sustainable resource management.

You should focus upon either a national or a local case study of resource conservation and discuss the successes and failures of strategies designed to manage one resource. China's hydroelectric power (HEP) scheme is a good example of a way in which it is trying to reduce its dependence on coal.

Key definitions

- **Ecological footprint**: The theoretical measurement of land and water a population requires to produce the resources it consumes and to absorb its waste under prevailing technology.
- **Sustainable resource management**: Conserving the quantity and quality of resources to maintain their supply for future generations.

Explain the relationship between energy usage and ecological footprint for **one or more** countries. *[5 marks]*

[Taken from paper 1, May 2011]

How do I approach the question?

This question is asking you to explain the relationship between energy usage and ecological footprint for one or more countries. You must first describe the relationship—countries with a higher energy usage generally have a higher ecological footprint. That will score you one mark. The explanation must link energy usage to ecological footprint.

One explanation that is very detailed could score four marks; two developed statements of explanations could also score four marks, as could four basic statements of explanation. If you do not refer to any countries the maximum mark you can score is three marks.

What are the key areas of the syllabus?

- The core: Patterns in resource consumption
- Sub-topic: Changing patterns of energy consumption

This answer achieved 1/5

The student does not describe the relationship between energy usage and ecological footprints. The statements are very generalized. More succinct explanation is needed.

Energy usage requires in different countries the building of infrastructure which one relates to the amount of earthspace we use in order to sustain our populations. On the UK for example the amount of space taken by the nuclear powerstation and the sea and soil pollution due to nuclear waste as an impact on the amount of earthspace used in other countries such as China the building of enormous dams to provide energy and resources used like water and coal from the mines has direct impact on the ecological footprint, the amount of earthspace used to sustain a population. [1]

Section B

Section B answers are in the form of short essays. As they account for a quarter of the marks on this paper, you should allocate about a quarter of the time available to you—about 24 minutes. The questions ask you to: "discuss the extent", "discuss the view" or "to what extent". You should aim to show at least two different points of view and be able to come to a conclusion, based on the evidence you have given in your answer.

To what extent do migrations bring benefits to both their origins and their destinations? *[15 marks]*
[Taken from paper 1, May 2011]

How do I approach the question?

This question asks you "To what extent do migrations bring benefits to both their origins and their destinations". You must cover both the origin and destination of migration. At the origin, benefits include remittances that are sent home, reduction in population pressure, and reduced demand for resources. At the destination, benefits include increased workforce, workers who are prepared to take low-paid and unpopular jobs. There must be an evaluation to reach markbands E and F.

What are the key areas of the syllabus?

- The core: Populations in transition
- Sub-topic: Movement responses—migration

This answer achieved 6/15

Migrations originated from push and pull factors, factors which determine the reasons why people move from some countries to others. If people move away from some countries with economic problems like unemployment the emigration of young workers will help to reduce the rate of unemployment. As the rate of unemployment decreases the rates of crime and violence will decrease as well. Workers emigrated from their own countries often still have their family in their origin countries therefore once they manages to find a job placement in a more developed country they send home money "remittances" this not only causes more money to circulate into the countries but also helps to increase the standards of living of the immigrated workers who send them valuable foreign currency.

An acceptable but very general answer. It would be beneficial if the student had used a case study to back up this point and/or developed the ideas of further economic benefits. For instance, would it benefit the entire population in the origin country or a more localized population in the migrant's home region?

Some migrants in case of economic success are also likely to return to the origin countries and reinvest money in those country creating further economic benefits for the entire population of the origin country.

This statement would benefit from some supporting facts and figures.

During migration benefits are evident also for developed countries which are usually the destination of migrant workers.

The student misses the opportunity to provide a list of the types of jobs that are undesirable or to give some figures, for example, the proportion of Mexicans working on farms in the USA.

Lack of clarity. The student is not really making sense here.

The essay needs to provide examples and evaluate more to score a higher mark.

With the presence of a new young workforce all the undesirable jobs in the destination countries are filled, we assist to the creation of multicultural societies like in the USA or the UK even sometimes there are episodes of racism.

Even if these are some disadvantages in migration for origin countries like the loss of young workforce and for the destination countries like the displacement of the local work enormous advantages are brought from migration to both origin countries, economy revitalised by remittances and destination countries which will have lots of young workers to employ in each section and who will sustain the country's economy in case the population decreases or in case there ore undesirable places to be filled. [6]

Examiner report

This answer provides some relevant information but development is limited and goes off the point. Limited evaluation—the question asks "to what extent", and this has not been answered.

7. Freshwater—issues and conflicts (option A)

This optional theme focuses on water on the land as a scarce resource. It considers the ways in which humans respond to the challenges of managing the quantity and quality of freshwater, as well as the consequences (whether intended or unintended, positive or negative) of management.

The theme includes both the physical geography of freshwater (basic hydrology, such as the water cycle and water balance, and floods) and human impacts on water quality (such as irrigation and agriculture, and wetland management).

This theme should include the study of **at least one** detailed case study of a named drainage basin. This is the area drained by one named river and its tributaries. Reference should be made to additional examples, at a range of scales, in less depth, wherever appropriate. These scales can be local, regional or international (such as the Nile and Mekong drainage basins, which include land from a variety of countries).

The water system

You should be able to:

- examine the inputs, outputs, stores and transfers of the hydrological cycle
- discuss the causes and consequences of the changing balance between water stored in oceans and ice.

You must be able to give a detailed account, including reasons and/or causes, of how the hydrological cycle works. This could be at any scale, from a local drainage basin hydrological cycle to the global hydrological cycle.

"Discuss" is a higher-order command term—you need to offer a balanced review that contains a range of factors or arguments. Conclusions should be supported by appropriate evidence.

You should be able to:

- explain the concept of maximum sustainable yield of freshwater in terms of a balance between inputs and outputs.

A sustainable yield can only be supported if the amount of water used is less than the difference between the inputs and the outputs (that is, excluding human extraction). It is often easier to see where the yield is unsustainable rather than where it is sustainable.

Key definitions

- **Maximum sustainable yield**: The maximum level of extraction of water that can be maintained indefinitely for a given area.

Drainage basins and flooding

You should be able to:

- examine the functioning of a drainage basin as an open system with inputs, outputs, transfers, stores and feedback loops.

"Examine" is a higher-order command term. Here, it requires you to consider the concept of a drainage basin in a way that investigates the interrelationship of stores, inputs and transfers.

You should be able to:

- define stream discharge
- examine its relationship to stream flow and channel shape.

"Define" means giving a precise meaning. Here, you should identify the units in which discharge is measured (normally cusecs or cumecs).

Remember that there are other factors that influence discharge, along with stream flow and channel shape.

You should be able to:

- describe the characteristics of a hydrograph
- examine the reasons for spatial and temporal (short-term and long-term) variations in hydrographs
- examine the role of hydrographs in forecasting the magnitude, spatial extent and timing of floods.

Upstream hydrographs can be used to predict the magnitude and timing of floods, as in the floods at Rockhampton caused by the Fitzroy River in Queensland, Australia in 2011.

A hydrograph relates the discharge of water in a river to a particular storm event or other events. Factors affecting hydrographs include climate, vegetation, basin size and shape and, increasingly, human activities.

Hydrographs are often used to show how a river is likely to react to a flood event.

You should be able to:

- discuss the natural and human causes and consequences of a specific river flood.

This requires a named flood event, for example, the floods in 2011 in Queensland, Australia. (The flooding of Rockhampton by the Fitzroy river would also work well here.) Both natural causes (monsoon rains, a tropical cyclone, prolonged heavy rain) and human causes (building on a floodplain) added to the impact of the floods.

Key definitions

- **Drainage basin**: The area drained by a river and its tributaries.
- **Drainage divide**: Also known as a watershed, it is the line defining the boundary of a river or stream drainage basin separating it from adjacent basin(s).

Management issues and strategies

You should be able to:

- examine the hydrological changes resulting from the construction of dams and reservoirs
- examine the costs and benefits of dams and reservoirs as part of multi-purpose schemes.

This sub-topic focuses on the hydrological changes (increased evaporation, decreased interception, regulated channel flow, changes in local precipitation) caused by the construction of dams and reservoirs.

Dams and reservoirs should be seen as part of a multi-purpose development scheme (including flood control, generation of hydroelectricity, improvements in navigation, increase in irrigation, the development of water sports and tourism, for example). A named case study would be beneficial.

You should be able to:

- explain the stream channel processes (erosion, transport, deposition) and explain the resultant landforms found on floodplains
- examine the human modifications of a floodplain and their effect on the size and probability of floods
- evaluate the costs and benefits of alternative stream management strategies.

This sub-topic covers well-known processes that you need to understand in detail, as you may have to explain them. The landforms are limited to those found on floodplains—meanders, cutoffs (ox-bows), levees, rapids (cataracts), and, of course, the floodplains themselves.

Human activity has increased the magnitude and speed of flooding by increasing both the amount of impermeable surface and the drainage density in built-up areas.

Alternative stream management schemes include levees, flood-relief channels, reservoirs and deepening channels, among others.

Make sure you do **not** write about waterfalls and gorges. These are not found on the floodplain.

You should be able to:

- explain the functioning and management of artesian basins and aquifers, distinguishing between natural and artificial recharge
- examine the environmental impacts of groundwater abstraction.

An aquifer is a water-bearing rock and an artesian basin is where water is confined geologically but is able to rise up to the surface under pressure. "Recharge" refers to the refilling of the basin or aquifer, either as a result of natural processes or by human activity.

Groundwater abstraction can lead to subsidence, declining water quality and, in coastal areas, saline intrusion.

You should be able to:

- describe the role of wetlands as a water resource
- evaluate the effectiveness of the management strategies that have been adopted in a major wetland.

Wetlands are an important supply of water. They act as a form of flood control and contain varied biodiversity. You should refer to a case study (such as the River Kissimmee in Florida or the River Brede in Denmark, where the natural rivers have been restored).

Land-use management, management of visitor numbers and removal of vegetation are all examples of management strategies that have been used in wetlands.

You should be able to:

- examine the environmental impact of agriculture and irrigation on water quality: salinization, agro-chemical run-off, the pollution of groundwater and the eutrophication of lakes, rivers and wetlands.

Different types of irrigation may have different effects: centre-pivot irrigation in a warm country is likely to cause more harm than drip irrigation; commercial farming is likely to cause more damage than subsistence farming. Agro-chemical fertilizers can include NPK fertilizers that are linked to eutrophication.

It is very easy to overgeneralize. To avoid this, give named and located examples to support the points you are making.

Key definitions

- **Wetlands**: Areas that are regularly saturated by surface water or groundwater, including freshwater marshes, swamps and bogs.

Competing demands for water

You should be able to:

- examine the competing demands for water in a specific river basin
- evaluate the strategies that have been adopted to meet these demands.

Competing demands include farming, industry, municipal (household), recreation and services. A named river basin is a requirement.

To "evaluate", you will need to consider a range of measures (for example, metering, reservoirs, inter-basin transfer) and consider which is best, and why.

You must include a named drainage basin in your answer, with actual strategies—a generalized answer will not score well here.

You should be able to:

- discuss an example of an international conflict related to freshwater.

Good examples include the Mekong, the Nile and the Jordan. The focus is on international conflict.

(a) Draw a labelled diagram to show the main features of an artesian basin. *[4 marks]*

(b) Explain the environmental impacts caused by groundwater abstraction. *[6 marks]*

(c) Evaluate the success of the management strategies used in **one** named wetland area. *[10 marks]*

[Taken from paper 2, May 2011]

How do I approach the question?

(a) Make sure that your diagram is large, carefully drawn and well labelled. Annotations and an explanation are not required.

(b) You should refer only to groundwater abstraction, not surface water abstraction, such as rivers and lakes.

(c) You will first need to name, locate and describe your chosen wetland and then explain why the management strategies are necessary. The nature of the strategies, and their mode of implementation, should follow, and this will vary depending on your choice of wetland. The evaluation is important and you should devote at least half of your answer to this. "Evaluate" means you should say how well the strategies have worked from several perspectives, such as conservation and economic costs. If relevant, you may also choose to comment on their limitations. If the strategies have only recently been implemented, you could predict their likely success (or failure) in the future.

What are the key areas of the syllabus?

- Option A: Freshwater—issues and conflicts
- Sub-topic: Management issues and strategies

This answer achieved 20/20

This is a good quality diagram showing a basin with the characteristic synclinal features. There is sufficient accurate information here to qualify for full marks.

(a)

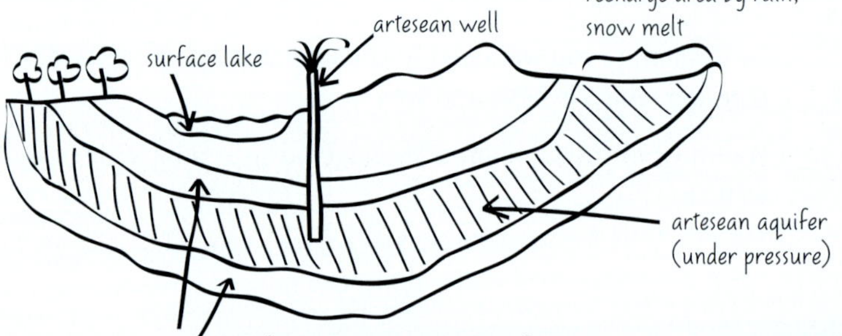

[4]

This is a comprehensive answer that covers the important aspects: saltwater incursion on coastlines leading to salinization and ground subsidence.

(b) Groundwater abstraction can cause many environmental impacts if not extracted at a sustainable rate. Groundwater extraction in the coastal [aquifer] in the Garza stripe, Israel causes sea water infiltration, causing [salinisation], making the water unsuitable for drinking and agriculture, killing citrus trees. Ground water extraction can also lower the ground water table and cause sinkholes and ground subsiding as seen in the Aral Sea region in Russia.

In addition, unregulated extraction of ground water can cause [salinisation] of soil and wastage through evaporation. Bore wells left flowing uncontrolled in the Great Artesean Basin result in [salinisation] of soil as water left on the surface evaporates. 208ml a day is lost due to unregulated bore wells. [6]

This is an excellent factual introduction.

(c) Wetlands are of great ecological importance as they serve many purposes in the ecosystem. They can act as carbon sinks (contains 1/5 of the worlds carbon), protect coastal area from storm surge as well as providing shelter and food for many species of fish and migratory birds. 50% of wetland are lost between 1990-2000, therefore conservation of wetland is very important.

There are two wetlands areas in Hong Kong that are very close together, and collectively grouped as the Deep Bay Wetland area. The 2 wetlands in the area are Mai Po and The Hong King Wetland Park (HKWP)

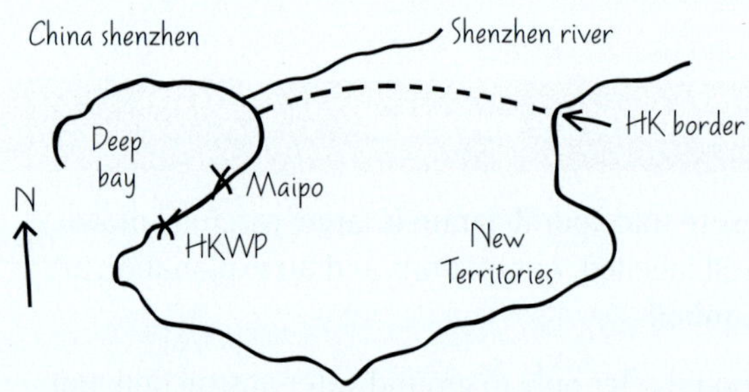

The Mai Po Wetland was declared as a RAMSAR site of environmental importance in 1987. Every season, 3000-5000 shore birds stops at Mai Po, including 22 endangered species especially the Black Face Spoon bill. The wetland area is jointly managed by the Agriculture and Fishene Department (AAFD) and the World Wildlife Fund (WWF). AAFD set up restricted access to the area only allowing 40000 visitors per year in order to minimise visitor impacts.

Alien species are frequently controlled and the water quality is constantly monitored. The wetland area organises many education tour and bird watching activities to enhance public awareness to wetland conservation. In fact Mai Po is also a site for wetland management training.

Next to Mai Po is the HKWP, constructed as an ecological buffer zone due to the development of the Tim Jnui Wai new town. The total area of the HKWP is 60 ha and it receives 100,000 visitors per year. There are no restricted areas but tall grass and warning signs are placed around the park in order to minimise visitor impact. Artificial nests are placed at site where there are little visitors for birds and bats. Water level of lakes and ponds are monitored and control by smaller resulting gates. However as this site is relatively new visitor impacts are not prominent yet.

In terms of attracting visitors both sites are doing very well, however Mai Po have limited areas which is very important to avoid too much disruption to the areas. Also, pollutions from the Shenzhen river is starting to affect the wetland areas. The number of water birds in Mai Po decreased to 3000. The Shenzhen river is outside HK border, thus the government have no control over the pollution. Furthermore there are signs of rising mudflats due to increase sediment and pollution from the Shenzhen river. This means that area where the birds stay is advancing out of HK legal border making it more difficult. To improve management, the HK government should seek cooperation with the Shenzhen authorities to regulate any river engineering schemes in the river (e.g. channelization) as well as monitoring industrial waste discharge. Without this, the management scheme will not be successful in the future [10]

Examiner report

Several strategies are discussed with references to wetlands in Hong Kong. Monitoring activities are described and their purpose explained in terms of ecological conservation. Detail is given about specific strategies, some of which are evaluated. The student shows excellent case study knowledge with a strong focus on the question throughout. Overall, it is well located, detailed, evaluative and so answers the question without having to evaluate all the management strategies used.

This answer achieved 7/20

The diagram shows an aquifer and the well, but omits essential features such as impermeable rock strata and the source of the water. The location of the aquifer in relation to these other features is essential and must be shown.

(a)

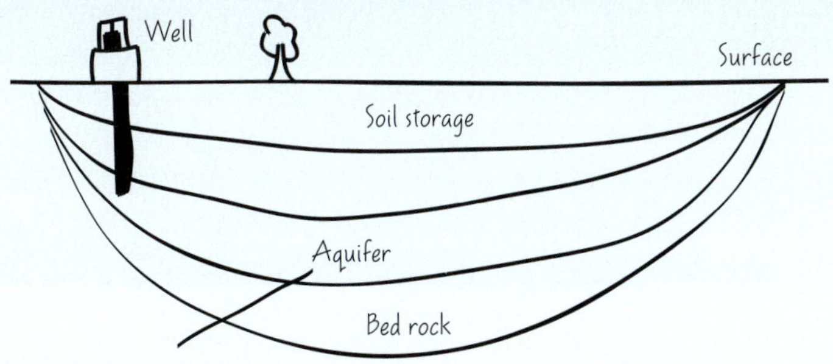

[1]

Salinization is correctly identified, but there is much irrelevant material that does not relate to groundwater.

(b) Several impacts caused by ground water abstraction is that there is the threat of small plants to die as they can't abstract moisture from the soil because they are arid and dry. However the major threat in ground water abstraction is the salinisation, agro chemical run off and eutrophication of the water in the aquifer. What this can provoke is that the water in the aquifer cannot be utilised on agriculture or simply undrinkable. Despite this another environmental impact for example what happened in the London Basin in the UK the community did so many wells that the water table decreased immensely that there is the threat of creating air pockets causing the surface to subside. Another environmental impact is the soil degrading, extracting the water which consists drilling holes in the land the soil structure can be extremely damaged therefore nutrients in the soil can be lost making the agriculture on land nearly impossible or making the soil erode. In conclusion ground water abstraction can impact the environment in many ways such as soil quality is affected, water quality is also affected through salinisation, agro chemical run off and eutrophication and finally biodiversity no nutrients in soil plants can't grow. [1]

This is a suitable example, but is inadequately described and located.

(c) A wetland is an area saturated by surface water or ground water. An example of a management strategies used in a wetland is the Norfolk Broads a shallow sheet of fresh water found in the eastern UK for multi purposes.

The Norfolk Broads is a wetland visited by many tourists which allows them to visit the incredible biodiversity of the ecosystems in many system of ways.

The student misunderstands the process of eutrophication and incorrectly links it to boat wash. Eutrophication is caused by nutrient overload from organic run-off from agricultural fertilizers and/or sewage. Wash from boats creates a different problem—that of bank erosion.

For this reason the Norfolk Broad is in great threat were recreational boating's used by tourists caused the river banks to erode and affect fauna by the wash of boats. Another environmental impact is the eutrophication of water by the wash of boats which consequently made grow a vast of algae's which interrupted sunlight entering in the water affecting fish. When the people got conscience of this problem several strategies were adopted to overcome the problems. One of the strategies involved was the use of pilings to not ultimately erode the banks this strategy was expensive though successful. Furthermore the government introduced two authorities to conserve the Broads:

Three relevant strategies are mentioned, but detail is lacking.

One controlling the recreational boating's

One control on water quality, sewerages and habitats.

Finally a last strategy to protect fauna particularly fish was the creation of refuges in the middle of the river to protect them from the wash of boats.

In the end all of these strategies where proven to be very successful as authorities on the broad managed control the happenings in the water. However the only downside was the environment was still in threat because certain tourist were trampling the area therefore destroying the environment. Anyhow the creation of two authorities and pilings in the banks and refuges in the river; even though it was an expensive process benefits are better in the long term, therefore the strategy was a very successful one. [5]

Examiner report

The student selects a valid case study, but their knowledge of the value of the Norfolk Broads is weak and the reasons for implementing management strategies are not fully explained.

This wetland should be clearly located, its physical attributes outlined and the tourist impacts described. The evaluation is very thin and the ecological/environmental success should be measured.

8. Oceans and their coastal margins (option B)

This optional theme is about oceans and their margins rather than the more traditional geographic approach to coastal landforms and processes.

Since oceans cover over 70% of the Earth's surface, oceans are of great importance to humans in a number of ways. This optional theme provides an introduction to the physical characteristics and processes of the oceans with particular reference to the atmosphere–ocean link, concentrating on the important role that oceans play in influencing climatic conditions. Increasingly, the role of oceans in the complex issue of climate is being recognized.

Issues arising from the oceans as resource bases are also considered. Some of these resources are being released as the climate changes and there is a scramble to utilize these resources.

The emphasis in the section on coastal margins is on management. Consequently, a detailed study of the physical characteristics and processes of coasts is not required, although a basic knowledge is essential for understanding management strategies.

You will need to provide a number of detailed examples or case studies in several sub-topics. If possible, these should include examples from areas that you know or have visited.

Introduction to oceans

You should be able to:

- describe the distribution of oceans and ocean currents.

The command term "describe" requires you to give a detailed account. You should know the major oceans, and some of the main ocean currents. You should try to learn the names of one hot and one cold current from the northern hemisphere and the southern hemisphere.

You should be able to:

- describe the main features of oceanic crust and ocean floor morphology
- explain the occurrence of oceanic volcanic features, trenches, transform faults, mid-ocean ridges and rifts in terms of plate margins.

You should be able to give a detailed account (composition, density and any other distinguishing feature) of oceanic crust and size, shape and other distinguishing features of landforms of the ocean floor (deep sea trench, mid-ocean ridge, rift valleys).

You may be asked to draw an annotated diagram to explain the formation of the listed features. You should be able to relate these features to different types of plate boundary (destructive and constructive plate margins).

You should be able to:

- describe the horizontal and vertical spatial variations in the temperature and salinity of ocean water.

You should be able to describe in detail how salinity levels vary with latitude and with depth of water. There are important differences between tropical and polar areas, as there are between shallow water and deep water. Seas off large rivers, such as the Amazon, may have lower salinity levels owing to the freshwater they receive.

Key definitions

- **Distribution—**where something is located. There may be a pattern to some distributions.
- **Morphology—**relating to the shape of a feature.
- **Ocean currents**: The major currents move huge masses of water over long distances.
- **Oceanic crust**: Crust that is relatively thin but dense. It increases with age away from a mid-oceanic ridge.
- **Warm ocean currents** move water away from the equator. In contrast, cold ocean currents move water away from cold regions towards the equator.

Oceans and climate

You should be able to:

- explain the thermal transfers of energy within oceans and the importance of oceanic conveyor belts.

This is a complex topic and requires a detailed account of the conveyor belt, including causes. Do not confuse this with normal ocean currents.

You should be able to:

- explain the atmosphere–oceanic interactions associated with ENSO (El Niño Southern Oscillation)
- explain the El Niño and La Niña phenomena and their climatic, environmental and economic effects.

Again, this is a complex topic. You may find it easier to learn by drawing annotated diagrams of the conditions associated with El Niño (for example, showing areas of high pressure, low pressure, rain and drought) as well as for La Niña.

Case studies of both El Niño and La Niña would be useful in explaining these phenomena—the floods in Australia in 2011 were blamed in part on La Niña. It is easy to confuse El Niño and La Niña—both are part of the Southern Oscillation.

You should be able to:

- examine the role of oceans as a store and source of carbon dioxide (CO_2).

Here, you should consider the interrelationships and feedbacks that arise as the oceans are both a store and a source of carbon.

You may previously have studied a model of the carbon cycle (in biology or ecology) in which the role of the oceans is minimal—the focus here is on the importance of the oceans with respect to carbon.

Key definitions

- **El Niño**: A reversal of normal weather conditions in the Pacific, that is, warmer water in the eastern Pacific and cooler water in the western Pacific leading to floods in South America and drought in parts of Australia. Its impact is linked to global changes in weather patterns.
- **La Niña**: An intensification of normal weather conditions leading to drought in Peru and flooding in the western Pacific (for example, the Queensland floods in Australia in 2011).
- **Oceanic conveyor belts**: A global thermohaline circulation, driven by the formation and sinking of deep water and responsible for the large flow of upper ocean water.

The value of oceans

You should be able to:

- identify the biotic and abiotic resources of continental shelves, oceans and ocean floor deposits.

Biotic (living) and abiotic (non-living) resources, such as manganese, sulphur, gold, oil and gas can be found in continental shelf areas and on the ocean floor. Some resources, such as fish, are in the oceans rather than on the ocean floor. Work out which resources are found in shallow (shelf) areas and which are deep sea resources.

Be aware that not all the resources are found everywhere in the oceans. Some are quite localized.

You should be able to:

- examine the spatial and temporal consequences of overfishing.

You need to consider where overfishing has occurred, and when it has occurred. What is the trend in overfishing? You should also look at the consequences of overfishing. These will be environmental (such as falling fish stocks), economic (falling yields, lower catches) and perhaps political (friction between governments over who has the right to fish).

You should be able to:

- evaluate a case study of a conservation policy implemented to provide sustainable fish yields
- describe the sources and distribution of pollution in the oceans.

To "evaluate" you must make an appraisal of the policy—what are its strengths and weaknesses? What are its impacts? Did it achieve its goals? An example of a good case study is the common fisheries policy of the EU, although other policies may also be appropriate. When you evaluate you must reach a decision. You should not just describe the policy but you should state what you believe to be its merits and its faults.

You should be able to:

- discuss the implications of the pollution of oceans by the disposal of radioactive material, oil and chemical waste.

To "discuss", you must offer a balanced review that contains a range of ideas or arguments. Opinions should be presented, and supported by facts and figures (evidence). You may get a question on one or more of these forms of waste—you need to study them all. You should include supporting evidence—the more evidence you provide, the better your answer will be. Answers that lack case studies are likely to score a low mark.

Key definitions

- **Abiotic**: Non-living materials, such as nutrients and rocks.
- **Biotic**: Living organisms, such as plants and animals.
- **Continental shelf**: Gently sloping (1°) margins of the oceans.
- **Ocean floor**: This refers to the deep ocean floor.
- **Overfishing**: This occurs when the rate of fishing is greater than the rate of natural replacement.

Geopolitics of oceans

You should be able to:

- discuss the sovereignty rights of nations in relation to territorial limits and exclusive economic zones (EEZ).

When you are asked to "discuss", you should offer several real-life examples. The UK and Iceland, the extension of territorial rights in the Arctic, Australia and East Timor are all good examples. Answering the question without providing suitable, detailed examples will only score low marks.

You should be able to:

- examine a geopolitical conflict in relation to an oceanic resource, other than fishing.

The command term "examine" suggests that you unravel some of the issues related to the conflict. This needs one or more case studies for impact—Rockall, the Falkland Islands, and the scramble for the Arctic are all good examples.

You should not cover fishing in your answer!

Key definitions

- **Exclusive economic zone (EEZ)**: An area in which a coastal nation has sovereign rights over all the economic resources of the sea, seabed and subsoil, extending up to 200 nautical miles from the coast.
- **Geopolitics**: This term means "geographical politics". It refers to the geographical relationships between different states. It also refers to the political power of nations, based on the economic resources they control.
- **Pollution**: The addition to an environment of a substance or an agent, such as oil, by human activity, at a rate greater than that at which it can be rendered harmless by the environment, and which has an appreciable effect on the organisms within the environment.
- **Sovereignty**: Having the right to rule and govern a defined territory.

Coastal margins

You should be able to:

- examine the relationship between coastal processes (tides, wave action, littoral drift, wind action), lithology, subaerial processes and different coastal landforms
- identify the major landforms of beaches, dunes and cliffs along advancing and retreating coasts.

"Examine" is a higher-order command term. You will need to look at the interrelationship of coastal processes, rock type, weathering, mass movement and coastal features. Make sure you are clear about advancing and retreating coasts.

This part of the syllabus requires four hours of study. Your teacher may be tempted to spend much longer on this part as it is the traditional core of a coasts-focused syllabus.

You should be able to:

- discuss the conflicts that arise from competing land uses and from attempts to manage coastal hazards (tsunamis and storm surges, erosion, cliff failure), pollution, habitat restoration and aquaculture.

This requires the use of detailed case studies for best effect. Who are the competing users? What are the conflicts? How can they be managed? How does the management of hazards, pollution, habitat restoration and/or aquaculture cause conflict. Who is the conflict between?

Without case studies your answer could sound very general—and will not score well.

You should be able to:

- describe the conflicting pressures on a particular coastline
- discuss the management strategies adopted to resolve these pressures and evaluate their effectiveness.

This requires the use of a detailed case study—Studland Beach in the UK, Miami Beach in the USA and Omaha Beach in New Zealand are three good case study examples. Attention should be paid to the conflict between users and the methods used to manage the conflict.

An attempt to assess the success of the management strategies should be made.

A common mistake is to use too large an area, for example, "Nigeria" or "Kenya" rather than a specific example, such as "Jolly Harbour, Antigua".

Key definitions

- **Advancing coasts**: Depositional coasts that are growing as a consequence of sediment deposit and/or the infill of coastal marshes. Advancing coasts may also arise from a negative change in sea level (sea level fall or uplift of land).
- **Lithology**: The study of rocks with respect to their physical, chemical and textural characteristics.
- **Littoral drift**: The movement of sediment along a coast by wave action; also called longshore drift.
- **Retreating coasts**: Coasts along which the dominant processes are erosional, resulting in the coastline moving inland. Retreating coasts may also be caused by a positive change in sea level (sea level rise or a fall in land level).
- **Subaerial**: Occurring or forming at the surface of the Earth.

Coral reefs and mangroves

You should be able to:

- examine the development and the environmental and economic value of coral reefs and mangrove swamps.

You need to be aware of the formation of different types of coral reefs, such as fringing, barrier and atoll. You also need to know the economic benefits that coral can bring, such as employment, tourism, and reducing the impact of hazards. In addition, there are environmental advantages of mangroves and coral, including maintaining biodiversity and the preservation of breeding grounds for fish.

It is not always possible to give an economic value to coral or mangrove. Some of the benefits, such as genetic diversity, do not have an economic value—they are, in fact, priceless.

You should be able to:

- examine the causes and consequences of the loss of coral reefs and mangrove swamps.

You may be asked to answer questions on coral reefs, mangrove swamps or both, so you should prepare both for the exam. You will need to cover both the causes (natural and human-induced) and consequences (social, economic, environmental) of the loss of coral reefs and mangrove swamps.

Key definitions

- **Atoll**: A coral reef surrounding a central lagoon. Atolls may have developed around volcanic islands that have submerged, or may have developed upwards as sea levels have risen, for example, Clipperton Island in the eastern Pacific and the Maldives.
- **Barrier reef**: A coral reef running parallel to the coastline and separated from the shoreline by an extensive lagoon system, for example, the Great Barrier Reef in Australia.
- **Coral**: A living organism (polyp) living in clear, tropical waters. Corals live in communities known as reefs. Corals secrete calcium carbonate and build a skeleton.
- **Coral bleaching**: The expulsion of the algae from the coral, causing it to lose colour and become bleached.
- **Fringing reef**: A coral reef attached to the coastline, for example, the fringing reef off the coast at Eilat, Israel.
- **Zooxanthellae**: Algae that live in a symbiotic relationship with coral and give it its colour.

The graph shows the percentage of coral reefs that have been lost since 1960.

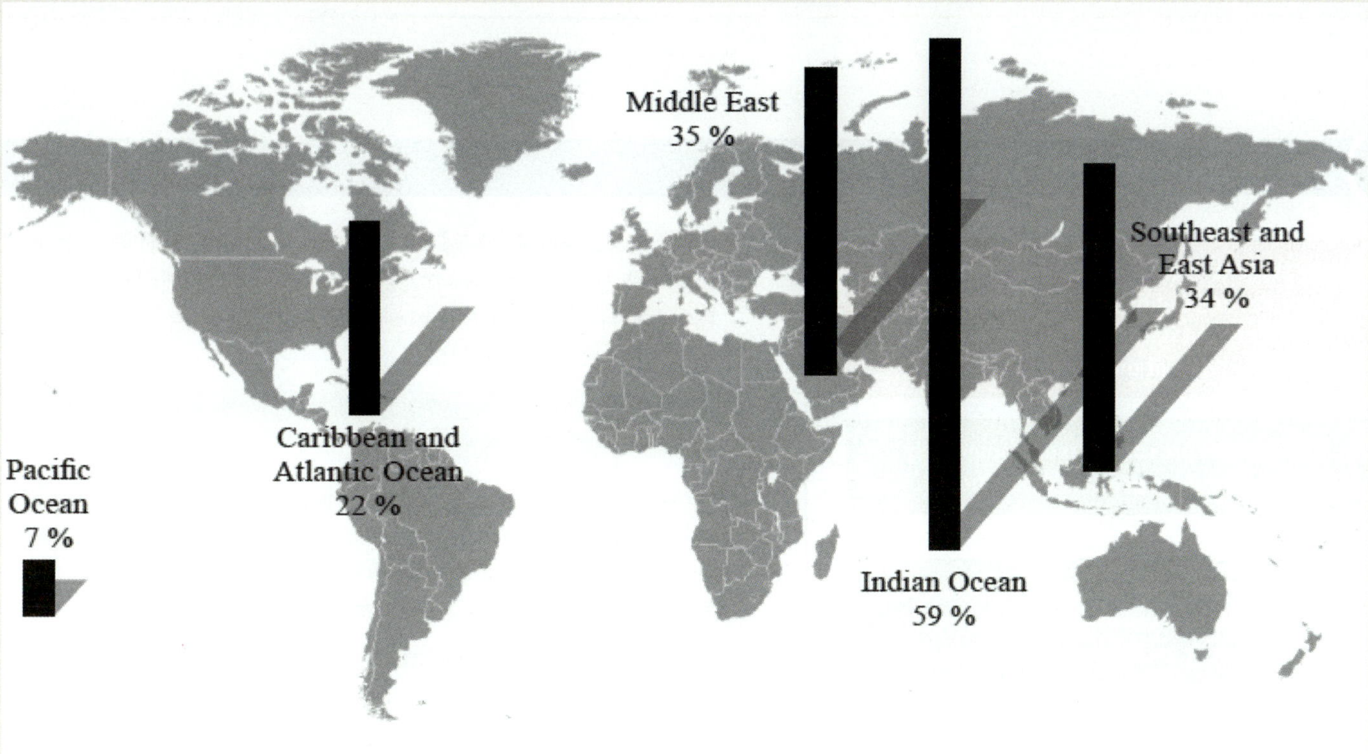

Source: adapted from http://commons.wikimedia.org/wiki/File:World_Blank_Map_(Mercator_projection).svg

(a) (i) Describe the pattern of loss of coral reefs. *[2 marks]*

(ii) State **two** physical factors which are needed for the development of coral reefs. *[2 marks]*

(b) Explain **three** benefits that coral reefs bring to people. *[3×2 marks]*

(c) Examine the conflicts which arise from competing land uses in coastal margins. *[10 marks]*

[Taken from paper 2, May 2011]

How do I approach the question?

(a) (i) Do not simply recite what is already shown. To achieve two marks, you will need to give an overview by describing the areas where loss is greatest and where it is least.

(ii) Choose **only** physical factors that cause coral reefs to start growing.

(b) Make sure that the benefits are distinct from each other. For example, biodiversity, coastal protection, tourism.

(c) The key word here is "conflicts". This needs careful definition in terms of incompatible land uses. There are numerous examples to choose from, such as the disturbance of wildlife caused by oil extraction, or the imposition of engineered sea defences in an aesthetic tourist resort. You should note other key words and phrases in the question to ensure that you answer it exactly. For example, you will need to examine (explain with justification) more than one pair of competing land uses and more than one coastal margin.

What are the key areas of the syllabus?

- Option B: Oceans and their coastal margins
- Sub-topic: Coral reefs and mangroves

This answer achieved 15/20

The student refers to the statistics and identifies areas of highest and lowest loss of coral reefs, but it is in the form of a list and there is no overview of the pattern.

(a) (i) The percentage of coral reef loss is highest in the Indian Ocean at a loss of 59%. The lowest loss of coral reefs is in the pacific ocean, with 7% loss. Southeast, East Asia and the Middle East have similar loss of 34% and 35% respectively. Apart from the Indian Ocean, loss of coral Reefs have been less than half (or 50%). [1]

Both answers are correct and appropriately stated without unnecessary description.

(ii) First physical factor is clear shallow water so sunlight can pass through. Second factor needed is some wave action to aerate the water. [2]

Coastal defence and tourist attraction are valid benefits and are well explained. However the second benefit, on fishing grounds, is too brief.

(b) The first benefit coral reefs bring to people is that they act as a natural coastal defence to coastal erosion and natural hazards. In Sri Lanka, 1km^2 of coral reef can reduce beach erosion by up to 20000 m^3 annually. Secondly, coral reefs are great fishing grounds as they are home to many fishes. The coral reef fishing industry is worth $2.5bn in South-East Asia.

Thirdly, coral reefs are very good tourist attractions. 90% of countries with coral reefs have established tourist industries, this contributes to a country's GDP and hence economic growth. An example is the Great Barrier Reef in Australia. [5]

The student recognizes the attraction of coastlines and the competing land uses that arise.

(c) Coastal margins tend to be more highly populated because of the more temperate weathers and in several countries, access to ports, trade, and hence Special Economic Zones, or urban centres. Over 60% of the worlds population live within 20km of a coast. With high population densities, there are bound to be many conflicts arising from competing land uses in coastal margins. This results in coastal squeeze.

This example is unsuitable.

Taking Presque Isle in Lake Erie in the USA as an example. Although Presque Isle is a sandpit, it is so huge that it faces the same marine processes and pressures. As Presque Isle has a very unique shape, and is home to so many species of birds, it is a very popular tourist destination. Below is a sketch of Presque Isle.

This diagram is poorly located and adds no credit. The perspective needs to be labelled (plan or profile).

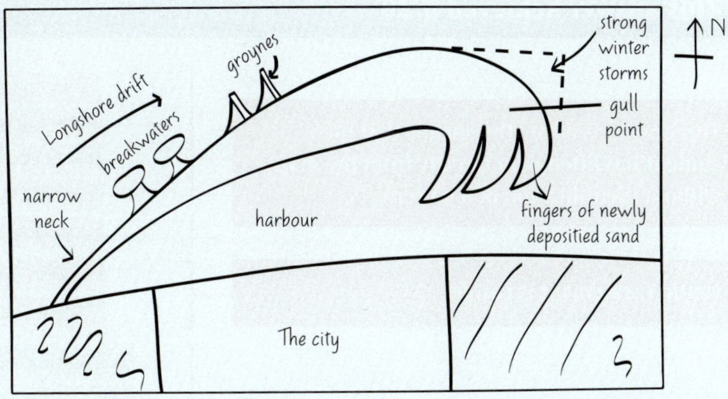

As can be seen, the neck of Presque Isle is very narrow and hence easily breached by winter storms. Presque Isle is naturally dry, so sand dunes are difficult to stabilize. With increased tourism, and dune activities, the sand dunes are destabilised and vegetation is often trampled on unconsciously. This makes the sediment on the coast looser and hence more vulnerable to marine erosion. The attempt to protect the neck from removal of sediment by longshore drift through the use of groynes starved Gull Point of Sediment, which was also bad as Gull Point faces erosion by strong winter storms, reflecting the conflict present.

This is not a marine environment.

Nuclear power stations also present competition for land use in coastal margins as they need a lot of water for cooling and are hence built close to sea. Take Dungeness in the UK as an example. Nuclear stations have radioactive wastes, and this puts people off living nearby for fear of health risks, thus many people have to be relocated, illustrating another conflict of land use in coastal margins – for settlement or for energy.

Nuclear power is mentioned and the conflict with settlement is clearly identified.

High population density in coastal areas also means increased demand for resources and for land for settlement. In south- east Asia, this has led to great destruction and deforestation of mangrove swamps.

Mangrove swamps are removed as their wood can be used for energy and the cleared land can be built upon. 40 years ago, Singapore used to be an island filled with mangrove swamps, but due to large-scale urbanisation, mangrove swamps have largely been removed leaving only a few in peripheral creeks and the Sungei Buloh nature reserve. However, mangrove swamps are great biological forerunners to more complex ecosystems, and their loss is severely felt. They can also reduce the maximum effect of tsunamis by up to 90% their intricate root system also provide unique breeding grounds, and are attractive to many young marine species. Prawn farming in South-East Asia has also led to the clearing of mangrove swamps in coastal margins. The combined competing pressures of urbanisation, prawn farming and settlement in coastal margins have led to the conflict of destroying mangrove reefs or keeping them to protect the environment and act as a natural coastal defence.

Useful and well-developed examples are given.

In conclusion, due to population pressures tourism, and urbanisation, coastal margins face increasing conflicts such as loss of mangrove swamps and increased marine erosion. [7]

Examiner report

The student introduces the topic well by identifying the universal attraction of coastlines in general and providing convincing factual evidence and justification for conflicts there. The example of Presque Isle is unsuitable because it is not strictly a coastal margin in accordance with the *Geography guide*, but the nature of conflict there is conceptually accurate and therefore deserves some credit. However, the case study in Singapore is entirely relevant, hence the recovery in marks.

The reason why this was awarded 7 marks (developed answer) rather than 9 or 10 marks (well-developed answer) was because the essay only has a short conclusion, with little evaluation of which conflicts might be the most significant in coastal areas, although there is a clear link made between population pressure and loss of mangroves and increased erosion. Although Presque Isle is a freshwater environment, there were examples that were clearly coastal margins.

This answer achieved 8/20

(a) (i) The most coral reef loss is found to be in the Indian Ocean, with a loss of 59%. Following is the Middle East with a loss of 35% and south East Asia with a loss of 34%. The Carribean and the Atlantic Ocean have lost 22% since 1960 and the Pacific Ocean has the lowest value of 7% loss since 1960. Overall LEDCs seem to exploit their coral reefs much more resulting to a greater percentage of coral reef loss since 1960. [1]

> This answer is a list. This means that the student has lifted data directly from the map without identification of a pattern.

(ii) 1 – waters must be rich biodiversity of fish
2- Still waters [0]

> Neither statement is correct.

(b) Coral reefs are important to humans for touristic aspects. For example Queensland in Australia has made a lot of profit due to the mass tourism which comes to visit the Great Barrier Reef.

Coral reefs also make sponges, and there is a great demand for souvenirs.

Coral reefs are the homes of around 50% of living fish therefore they offer a wide Biodiversity.

They also reduce CO_2 levels and keep the water clean. [4]

> Three benefits are correctly identified, but more explanation is needed throughout.

(c) Mangroves are trees that are found in the coastal tropical margins. For example Brazil has a lot of mangroves growing. Mangroves are in great importance because they offer medicine, food, and keep soil very fertile. However there is a competing land usage in coastal margins for many different reasons. Mangroves have become deforested for agricultural reasons, or for shrimp and fish production. However like this there is soil erosion– and loss of soil fertility which could eventually lead to desertification. Mangroves are also cut down to create new roads and transportation facilities. In this case the run off of the road flows into the oceans as there is no longer any protection which also affects coral reefs land competition in coastal margins also include housing opportunities. Not only is the population increasing and there is a greater demand for housing, but coastal land is mostly known to be more expensive luxury land. Therefore like this the income of a country can be increased. Other reasons include other agricultural possibilities such as a replacement for cattle. Like this more land is available for farming. Overall there are a lot of competing land uses seen by the removal of mangroves which might profit a country economically however, it definitely [ruines] the biodiversity, affecting land, fauna, animals and even the Oceans, and coral reef. [3]

> A confused set of ideas.

Examiner report

This is the answer to another question on the value of mangrove swamps and the reasons for their destruction. The student neither defines nor explains the issue of conflict between coastal land uses. The use of examples is very limited. Overall, this answer fails to address the key issues of the question.

9. Extreme environments (option C)

The extreme environments covered in this optional theme include cold and high-altitude environments (polar, glacial areas, periglacial areas; high mountains in non-tropical latitudes), and hot arid environments (hot deserts and semi-arid areas).

The physical background covered at the start of the optional theme provides the basis to understanding the processes and features that characterize these distinctive landscapes of rugged terrain, harsh climates and unproductive soils. Such conditions had limited settlement and economic activity in the past but, more recently, new technology has allowed the resource potential of these regions to be realized. Activities such as mining, oil drilling, irrigated agriculture and tourism make significant contributions to their economies.

The issue of maintaining sustainability is common to all extreme environments because they share the same problem of low carrying capacity. Their fragile landscapes and ecosystems are sensitive to change and recover slowly. Human activity of all kinds can have serious negative impacts, and the careful management of these environments and their resources is essential.

Challenging environments

You should be able to:

- explain the global distribution of each of the two kinds of extreme environment
- describe the relief and climatic characteristics that make these environments extreme
- explain how these characteristics present challenges for resource development and human habitation.

The two groups are: cold and high-altitude environments, and hot, arid environments. You are expected to recognize these environments on a world map.

You should know the landscape characteristics and seasonal climatic regimes of cold and high-altitude environments, which are variable.

The universal challenge is that of temperature, which restricts human habitation and economic activities, especially agriculture.

You should be able to:

- explain the other factors responsible for a low density of population in these areas: human discomfort, inaccessibility, remoteness
- identify the ways in which people adapt their activities to extremes of weather and climate.

You should choose case studies of one cold and high-altitude environment and one hot, arid environment, then, referring to each of the three factors, explain their low population densities.

Traditional adaptations to extremes of weather and climate include: migration of pastoralists in pursuit of water/warmth (transhumance); mitigation techniques to reduce winter damage, salinization, drought and flooding; modification of diet and food storage.

Key definitions

- **Arid**: Arid areas have less than 250 mm of precipitation each year.
- **Continentality**: Continentality is typical of large land masses, such as Asia, where, being far from the coastline, conditions are usually much drier and temperatures much more extreme.
- **Periglacial**: This term describes areas of snow and ice cover that are usually associated with permafrost or ground that remains frozen all year.
- **Semi-arid**: Semi-arid areas have 250–500 mm of precipitation each year and may be found in both tropical and temperate zones.

The physical characteristics of extreme environments

In this optional theme, you will need to show an understanding of the development of physical landscapes associated with three different extreme environments: glacial, periglacial and arid. You should be able to explain the physical processes involved in the formation of these contrasting environments.

In part (a) of the exam question, you may be asked to identify features on a photograph or a map (you should refer to chapter 5, “Data response skills”, for advice on these skills).

In part (b) of the exam question, typically allocated six marks, you may be asked to describe the characteristics and formation of a specific landscape feature. An annotated diagram may be required in the question. Even if this is not the case, you should be prepared to add an annotated diagram because it is likely to enhance your answer.

You should be able to:

- explain the advance and retreat of glaciers and the main features resulting from the processes of erosion and deposition by glaciers.

A glacial regime may be illustrated with a diagram showing the inputs (snow and ice) and the outputs (ablation and sublimation). This should include annotations giving an explanation of these processes leading to advance and retreat respectively.

You should also be able to explain the development of both erosional and depositional features and the role of ice in eroding, transporting and depositing materials.

You should be able to:

- explain permafrost, patterned ground, solifluction, thermokarst and pingos.

You should be able to:

- explain weathering and the processes involved in wind- and water-formed features
- explain the occurrence of flash flood features.

Key definitions

- **Arête**: Sharp edged ridge formed by freeze thaw weathering.
- **Arêtes and pyramidal peaks**: These are caused by the headward recession (cutting back) of two or more cirques.
- **Butte**: An isolated peak, often the remnant of a former mesa or plateau.
- **Canyons**: Very deep gorges, usually dry. If there is a river, it is usually exotic. The Fish River Canyon in Namibia is an excellent example.
- **Cirque (or corrie)**: This is a hollow found in mountainous glacial areas resulting from the accumulation of ice and snow, the deepening of the hollow followed by the processes of freeze, thaw and abrasion by ice.
- **Deflation**: The progressive removal of small materials leaving behind larger materials. This forms a stony desert or reg. In some cases, deflation may remove sand to form a deflation hollow. One of the best known is the Qattara Depression in Egypt, which reaches a depth of over 130 m below sea level.
- **Drumlins**: These are small, oval mounds up to 1.5 km long and 100 m high, which are deposited owing to friction between glacial ice and the underlying geology, causing the glacier to drop its load. The glacier streamlines the mounds as it continues to advance.
- **Erratics**: These are large boulders that are foreign to the local geology, for example, the Madison Boulder in New Hampshire, which is estimated to weigh over 4,600 tonnes!
- **Inselbergs**: Also known as **kopjes** or **bornhardts**, these are isolated domes of resistant rock. They can be formed by the parallel retreat of slopes (pediplanation) or by exhumation, where the basal surface is exposed as a result of surface stripping.
- **Lateral moraines**: The ridges of materials found on the edge of a glacier. Where two glaciers merge and the two touching lateral moraines flow in the middle of the enlarged glacier, they are known as **medial moraines**.
- **Mesa**: A relatively large area of plateau that has become isolated from the main plateau either by erosion or through slope retreat.
- **Moraines** are lines of loose rocks, weathered from the valley sides and carried by the glaciers. At the snout of the glacier is a crescent-shaped mound of **terminal moraine**.
- **Oasis**: An oasis is formed when water-bearing rocks are exposed at the surface. This is commonly achieved by deflation.
- **Patterned ground**: This is a general term describing the stone circles, polygons and stripes that are found in soils subjected to intense frost action.
- **Pediments**: Shallow slopes formed at the base of a cliff or steep hill.
- **Permafrost**: Permanently frozen ground (for at least two years). Approximately 20% of the world's surface is underlain by permafrost, which can be up to 700 m deep in places. Above the permafrost is the active layer, a highly mobile layer that seasonally thaws out and is associated with intense mass movements.
- **Pingo**: A pingo is an isolated, conical hill up to 90 m high and 800 m wide, which can only develop in periglacial areas. Pingos form as a result of the movement and freezing of water under pressure.
- **Plateau**: A large, flat area at high altitude.
- **Pyramidal peak**: Convergence of three of more arêtes formed by the headward recession of cirques and freeze thaw weathering.
- **Sand dunes**: An accumulation of sand forming regular features/landforms. These are formed where wind speed is high, constant and there is a large supply of sand.
- **Solifluction**: This term literally means "flowing soil". In winter, water freezes in the soil causing expansion of the soil and segregation of individual soil particles. In spring, the ice melts and the water flows downhill, carrying soil with it.
- **Thermokarst**: This term refers to subsidence caused by the melting of permafrost. This may be because of broad climatic changes or local environment changes.
- **U-shaped valleys**: These are formed where glaciers inroad original river valleys and their tributaries. Other features, such as **hanging valleys** and **truncated spurs**, can be found high up on their sides.
- **Wadis**: Dry gullies that have been eroded by flash floods. Heavy rainstorms create rushing torrents on steep slopes.
- **Yardangs** and **zeugens**: Wind-eroded landforms where the softer rock strata are removed, leaving the more resistant layers to form either long ridges (yardangs), where the strata are vertical, or mushroom-shaped features (zeugens), where the strata are horizontal. These ridges could be as high as 100 m and may stretch for many kilometres.

You should be able to:

- Hot arid areas: examine the opportunities for agriculture, the distinction between aridity and infertility, the importance of irrigation and risk of salinization, and the processes and factors involved in desertification.

Opportunities have optimistic outcomes, whereas challenges are negative prospects that are likely to bring problems. Opportunities may not be fulfilled and challenges may or may not be overcome. They are useful terms that allow for a wide range of outcomes. You should be able to recognize both the potential of, and the difficulties imposed on, farmers in these arid environments, and the human pressures that exacerbate the problems. Desertification is a global concern and you should refer to at least one case study. Remember that desertification is not confined to locations such as East Africa and there are valid examples in the richer world, such as in Mediterranean Europe.

You should be able to:

- Periglacial areas: examine the opportunities and challenges posed by permafrost and other characteristics of periglacial areas for resource development (mineral extraction and any associated settlement and communications)
- Hot, arid areas: examine the opportunities and challenges posed for resource development (mineral extraction and any associated settlement and communications).

In general, mining activity has been highly productive in both areas and you should be prepared for a question covering periglacial areas and hot, arid areas. One of the most efficient ways of answering such a question is to approach the issues and concepts through one or more case studies, for example, oil extraction in northern Alaska and uranium mining in Niger, rather than examining them in sequence.

You should be able to:

- examine the opportunities and challenges posed by the development of tourism and any associated settlement and communications in one type of extreme environment
- examine the impacts of tourism on the environment, such as mass movement and erosion, land degradation, vulnerability to hazards, aesthetic change, water usage and waste disposal.

The popularity of extreme environments for tourism has grown owing to tourists' endless search for novel experiences and remote destinations. The opportunities are mainly economic, but may also be social and political.

The impacts listed are environmental and emphasize the fragility of these extreme environments. Although some examples of the impacts are given, other, non-environmental challenges, such as economic leakage or cultural dilution can be included. The command term "examine" requires you to consider the argument or concept in order to uncover the assumptions and interrelationships of the issue.

You should be able to:

- discuss the degree to which human activities in extreme environments are unsustainable.

You should be able to define sustainability. You will not be expected to give the exact UN definition in the exam, but you do need to identify the principles of environmental management that conserve resources, avoid pollution and maintain productivity for future generations. The command term "discuss" means you should provide a considered and balanced review. You should look at both sides of the argument and draw a well-supported conclusion. The phrase "the degree to which" clearly indicates this approach. A likely view is that although human activities, for example, mining and tourism, are threatening these fragile environments, there is plenty of evidence for more sustainable practices being adopted. A good example would be that of game/safari tourism in Namibia.

You should be able to:

- discuss the potential impact of global climatic change (global warming) on the indigenous populations, settlement and economic activities in extreme environments.

Your answer should include two or more populations living in contrasting extreme environments that are affected by climate change in different ways. In high-altitude environments, such as in the Himalayas, glacial melt is a threat to countries downstream; in hot and arid areas, the drought/flood cycle may become more intense and more frequent. Indigenous populations may be forced to migrate to avoid drought and food shortage, and desertification may be a consequence of these impacts at their destination.

Key definitions

- **Desertification**: The expansion of the desert resulting from physical factors, such as drought, and human factors, such as overgrazing, overcultivation and poor land management. It is the final stage in the process of **land degradation**.
- **Land degradation**: The first stage of desertification, which includes several aspects of deterioration of the land, such as loss of organic matter in the soil, leaching of nutrients, loss of soil structure and the formation of deep gullies. It is typical of semi-arid areas that are exposed to drought/flood cycles.
- **Salinization** may occur in areas where annual precipitation is less than 250 mm per annum and where evaporation rates are high. Salts are drawn upwards through the soil by capillary action and are deposited in a crust on, or close to, the surface of the ground. Saline soils are toxic to many plants and are, therefore, unproductive.

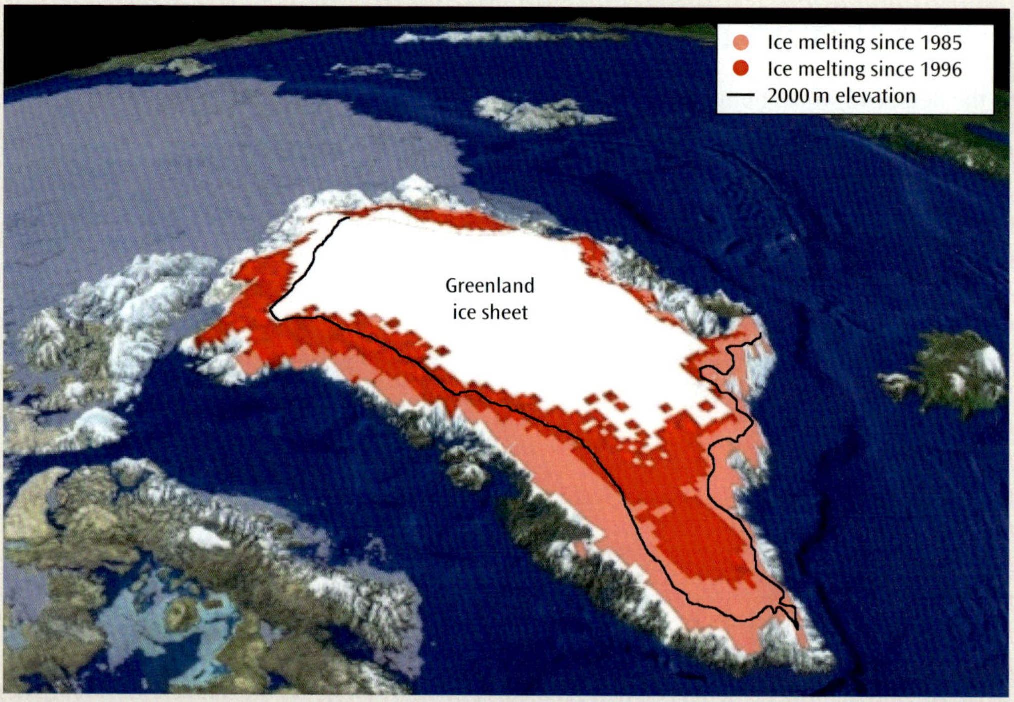

Source: Koni Steffen, CIRES/University of Colorado

The image shows changes in the melting of the Greenland ice sheet.

(a) Describe the changes that have taken place in the Greenland ice sheet since 1985. *[4 marks]*

(b) Explain how and why glaciers and ice masses have sometimes advanced. *[3+3 marks]*

(c) "Agriculture in hot, arid areas inevitably results in desertification." Discuss this statement. *[10 marks]*

[Taken from paper 2, May 2011]

How do I approach the question?

(a) Describe the changes in detail, making sure that you quote distances and directions. Give an overview of changes at the end.

(b) You will need to explain **how** glaciers and ice masses advance, where an accumulation of snow and ice exceeds melting. Explaining **why** requires a more holistic view of possible cause of climatic cooling.

(c) The focus is on agriculture in hot, arid areas, which allows for a broad interpretation in terms of location. You should include semi-arid areas, too, because desertification is particularly relevant here. (You cannot desertify an arid area that is already a desert.) You will need to explain the process of desertification and discuss how much it is controlled by human and physical factors. The best answers will look at both factors. Although the question does not mention examples, you must give some.

What are the key areas of the syllabus?

- Option C: Extreme environments
- Sub-topics: The physical characteristics of extreme environments; Opportunities and challenges for management

This answer achieved 18/20

The answer is good and recognizes changes over time and in space, giving precise locations. However, failure to quantify, or refer to the 2,000 metre contour and east–west differences means it does not achieve full marks.

(a) There has been an overall tincrease of ice melting in the Greenland ice sheet since 1985. In 1996, the amount of ice melting appears to have at least doubled since 1985. The ice melting in 1985 was largely closer to the coasts around the perimeter of the ice sheet, however, the increase melting in ice has meant that ice more inland has also started to melt. The ice melt is progressing inwards towards the centre of the ice sheet. However, the most inland areas of the Greenland ice sheet have yet to begin actual melting. The melting also looks to be particularly bad in the South of the ice sheet, as that is where melting has increased the most since 1985. [3]

The student gives a good explanation for expansion, and relates this to the balance between ablation and accumulation. There are also valid references to past ice ages.

(b) Glaciers and ice masses advance when ablation is less than accumulation. Also, temperatures make a difference, for example, glaciers and ice masses advanced during the Pleistocene Ice Age. Cold temperatures are more conductive for glacier advancement fluctuating temperatures also allow for glaciers to expand; if temperatures fluctuate just above and below freezing. This allows for glacier development as meltwater in warmer temperatures can fill gaps in ice and when they freeze again, becomes closer to forming glaciers. Heavy snowfall leading to increase in the glacier budget in the zone of accumulation will lead to glaciers advancing.

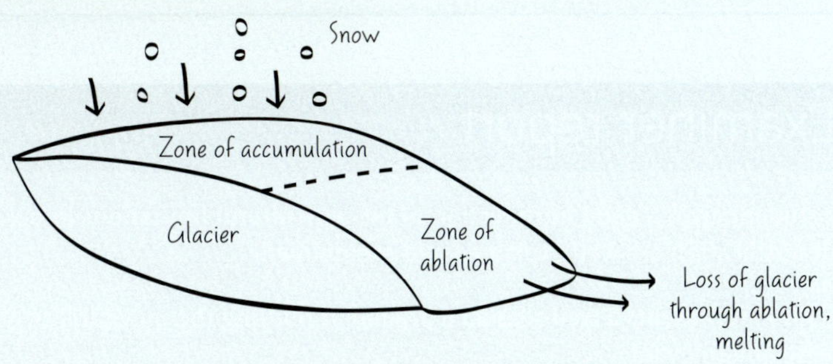

If accumulation is greater than ablation, the glacier will advance; otherwise, it will retreat.

Low global temperatures will allow for ice masses to advance because more areas will face temperatures that are below freezing, and hence more ice masses will develop. This is to support the statement on the Pleistoscene Ice Age mentioned above; during the ice age, glaciers and ice masses were seen to have advanced. [6]

Adverse physical conditions are well explained.

(c) Hot, arid areas are places with high average temperatures and low precipitation rates of below 350MM per annum. Since hot arid and receive little precipitation, there is low moisture, and soils are often hard-baked and dry. Agriculture in areas with already low moisture are bound to exert pressures, however, not all agricultural practices in hot areas are unsustainable or lead to desertification.

The student gives a detailed review of the human causes of desertification.

Agriculture in hot arid areas will lead to desertification only if inappropriate farming techniques are employed. For example, growing of cash crops such as cotton near the Aral sea was highly water demanding, and also the indiscriminate use of water led to greater areas at risk of desertification and the Aral sea shrinking to 10% of its original size. Cash crops take up a lot of nutrients and hence deplete the soil of nutrients quickly. Growing of monocultures also leads to risk of desertification. In many LEDCs short term financial goals are more important, and hence fallow periods are ignored. This poor farming practice will lead to destruction of soil structure since soil loses all their nutrients, and will thus become more prone to desertification. The growing of cash crops such as Tobacco are not fit for consumption, hence many locals have to farm on other lands for sustenance. Too often , since fertile lands are dedicated to growing cash crops, these farmers move on to marginal lands already at high risk of desertification. Agriculture on those lands is unsustainable and are very likely to lead to desertification. The high rates of potential evapotranspiration also means more water needs to be applied to crops in hot arid areas. If this leads to over abstraction of ground water from already low water tables, then desertification will be the result.

On the other hand, agriculture can be sustainable in hot arid areas. The growing of xerophytes such as olives will be a much more sustainable use of land in hot arid areas. Halophytes can grow in saline environments, so they will flourish in areas with low rates of precipitation and high rates of evaporation where soil tends to be more saline. The use of trickle irrigation will also be good in hot arid areas as they up to 90% more efficient than other irrigation practices. This will lead to less abstraction of ground water and lower risk of salinisation. Furthermore, contour ploughing or terracing will protect slopes from the infrequent but destructive flash floods that occur in hot arid areas. In fact, agriculture involving planting of marram grass hedges and shelter belt of trees will protect the hot arid environments from desertification. This is because the vegetation will stabilise sand dunes and loose soil with their binding roots. The application of gypsum during agriculture to fertilise soils and replace sodium ions with calcium colloids will also improve soil quality and prevent desertification. In conclusion, I disagree with the statement "agriculture in hot, arid areas inevitably results in desertification". Risk of desertification in hot and areas is higher, and inappropriate farming techniques will indeed lead to desertification, but if farmers follow fallow periods and practice good farming techniques, agriculture in hot arid areas can be sustainable, in fact, the vegetation might even help bind soil and lower the risk of desertification as top soil will be binded and not lost as easily. [9]

Examiner report

The student gives a good introduction to the agricultural limitations imposed by aridity. Processes such as poor management leading to desertification are well explained and exemplified.

The student presents a well-reasoned argument refuting the inevitability of desertification with good factual support. The conclusion is concise and relevant.

This answer achieved 11/20

The student correctly describes the melting of the ice sheet and locates features accurately enough, despite comments such as "left" and "right".

(a) In 1985, some but not all of the ice melting was above the 2000m line. In fact, most of the melting was below the line, which is especially true on the left and right sides of Greenland as seen on the photo (no compass is provided). By 1996 a significant amount of increased melting was taking place. In no area has the melting decreased, it has only increased in already melting areas and begun in areas not affected in 1985. [2]

The student shows a good appreciation of the process of advance and its relationship to inputs (accumulation) and outputs (ablation), but background reasons for cooling such as orbit and sunspot changes are needed.

(b) A glacier can be seen as a system. There is an accumulation zone, a processing zone, and an ablation zone. In the accumulation zone, generally in a cirque; ice, snow avalanches and debris builds up and forms the glacier over a period of time. The ablation zone is around the moraine, where meltwater and evaporation takes place. When the accumulation is greater than the ablation, the glacier grows and advances. If the level of ablation and accumulation are equal then the glacier size and position remains consistent. The glacier shrinks if ablation is greater than accumulation. Hence, glacier advance when the accumulation (input) is greater than the ablation (outputs). [3]

No physical factors are explained.

A simplistic explanation of human causes of desertification. More detail is needed.

These are all valid precautions against desertification.

(c) Agriculture in hot, arid areas may lead to desertification, but does not always do so. Generally, with modern technologies, desertification only really takes place when improper irrigation techniques are used.

Desertification is soil degration to a very serious extent resulting in the area turning in to desert like conditions, and soil degraded to the likes of sand. This is often caused but nomadic populations becoming sedimentary, and staying in one place. This often leads to over grazing and exploitation of the soil which previously was only subject to seasonal agriculture techniques are used as when the population was nomadic, as settled and more frequent agriculture requires different crops, irrigation techniques, protection, and fertilizers.

Agriculture in Arid, hot environments can be sustained and desertification avoided. To do so, proper water conservation techniques must be used, as water supply is very limited. Also, drought tolerant crops are much more likely to last in such an environment, and hence less likely to cause desertification to the environment as less soil stress is caused by replanting and less water is used.

Using mechanical soil protection measures are vital to prevent desertification. Windbreaks are required to protect the soil from wind erosion. This can also be done by enclosing the land which is to be farmed. Proper irrigation for the conservation of water and natural fertilizers such as manure can enrichen the soil and prevent desertification.

Nomadic agriculture tendencies have been on going for ages, such as the Sahel population. As long as the population are seasonally migrating, overgrazing and over working does not occur and hence desertification is avoided. Thus, the statement claiming that all agriculture in hot arid areas inevitably leads to desertification cannot be considered correct, as examples exist of agriculture which avoids it. [6]

Examiner report

The student only partially defines the process of desertification and neither addresses its global location nor the physical influence of drought.

Explanation of human causes makes some valid points regarding the increasingly sedentary lifestyle of nomadic populations. Some suitable remedial measures are mentioned, but reference to examples is very thin. The conclusion is relevant but too brief and needs to reflect upon the inevitability of desertification in terms of both space and time.

10. Hazards and disasters—risk assessment and response (option D)

Natural hazards exist at the interface between physical geography and human geography. The effects of natural hazard events are often intensified by human actions, although equally, human-induced hazard events are also affected by natural environmental conditions. The principles involved in studying natural hazard events are identical to those involved in studying human-induced hazards.

The focus of this optional theme is on the full range of human adjustments and responses to hazards and disasters at a variety of scales—local, national and international. The term "natural disaster" is deliberately avoided in this optional theme because it is not considered to be an accurate reflection of the multitude of underlying reasons that expose people to risk and subsequently create the pre-conditions necessary for a disaster to occur. Some people are more vulnerable to hazard events than others. Factors such as poverty and powerlessness increase a person's risk.

In studying this theme, you are expected to examine the following **four** hazards.

- **Either** earthquakes **or** volcanoes. Recent examples include Haiti in 2010 and Chaiten, Chile in 2010.
- Hurricanes (tropical cyclones, typhoons). Recent examples include Myanmar in 2008.
- Droughts. Recent examples include East Africa from 2000 onwards.
- Any one recent human-induced (technological) hazard resulting in an explosion or escape of hazardous material. Recent examples include pollution of the Qingshuixi River through Harbin, China. Although the Chernobyl explosion and the explosion at Union Carbide's factory in Bhopal occurred in 1986 and 1984 respectively, the impacts are still ongoing and so these can be used to answer some hazard questions.

These four hazards do not necessarily require an equal allocation of time; the precise balance will vary according to local preferences. The syllabus is designed to allow for flexibility but it is recommended that the overall approach should be concept by concept (such as vulnerability, risk and risk assessment), rather than entirely thematic (hazard by hazard). At least **one** case study of a hazard event (or disaster) is required for each of the four hazard types.

Characteristics of hazards

You should be able to:

- explain the characteristics and spatial distribution of the following hazards.
  - Either earthquakes or volcanoes
  - Hurricanes (tropical cyclones, typhoons)
  - Droughts
  - Any one recent human-induced (technological) hazard (explosion or escape of hazardous material)
- distinguish between the chosen hazards in terms of their spatial extent, predictability, frequency, magnitude, duration, speed of onset and effects.

You are expected to give a detailed account including reasons. Characteristics include magnitude, temporal occurrence, regularity, size and impacts. "Spatial distribution" refers to where they are found. A good example of a recent technological hazard is the pollution on the Yangtze River or the Qingshuixi in Harbin, China.

Key definitions

- **Characteristics of hazards**: Their spatial extent, predictability, frequency, magnitude, duration, speed of onset and effects.
- **Disaster**: A major hazard event that causes widespread disruption to a community or region that the affected community is unable to deal with adequately without outside help.
- **Hazard**: A threat (whether natural or human) that has the potential to cause loss of life, injury, property damage, socio-economic disruption or environmental degradation.
- **Hazard event**: The occurrence (realization) of a hazard, the effects of which change demographic, economic and/ or environmental conditions.

Vulnerability

You should be able to:

- explain the reasons why people live in hazardous areas.

People may live in hazardous areas for perceived benefits (such as rich volcanic soils), or because they do not perceive it to be a hazard, or because they have no choice in where they live, for example, poor labourers living close to the Union Carbide factory in Bhopal.

You should be able to:

- discuss vulnerability as a function of demographic and socio-economic factors, and of a community's preparedness and ability to deal with a hazard event when it occurs
- explain the reasons for some sectors of a population being more vulnerable than others.

Hazards are socially selective. Those who are well off can generally live in safe, comfortable environments. In contrast, the poor are often forced to live in areas that are unsafe, for example, on steep slopes or in low-lying coastal areas.

Some marginalized populations are forced into extreme environments where the risk of natural hazards is increased.

Key definitions

- **Demographic**: Relating to population characteristics.
- **Socio-economic**: The combination of social factors (including demographic, cultural and political) and economic factors.
- **Vulnerability**: The susceptibility of a community to a hazard or to the impacts of a hazard event.

Risk and risk assessment

You should be able to:

- examine the relationships between the degree of risk posed by a hazard and the probability of a hazard event occurring, the predicted losses and a community's preparedness for it
- explain the reasons why individuals and communities often underestimate the probability of hazard events occurring
- discuss the factors that determine an individual's perception of the risk posed by hazards.

The perception of a hazard is affected by experience. For example, a dormant volcano, such as the Soufrière Hills volcano on Montserrat pre-1995, poses limited risk. However, some hazards are regular, such as tropical cyclones. For this hazard, there is a regular hazard season.

Some populations are very mobile—recent migrants into an area may well be unaware of the potential hazards of that area.

An individual's response will be affected by their personality (risk-taker, security-conscious), what they can afford (preventative measures), where they live and levels of knowledge.

You should be able to:

- examine the methods used to make estimates (predictions) of the probability (in time and space) of hazard events occurring, and of their potential impact on lives and property
- discuss these methods by examining case studies relating to **two** different hazard types.

Estimates of where and when a hazard will impact are useful in that they allow a population or government agencies to plan for the impact.

You are required to study in detail **two** case studies for different hazards. Hurricanes tend to occur in certain locations (tropical and sub-tropical) in a set season whereas the prediction of volcanoes requires monitoring of their size and of the chemicals being emitted.

Key definitions

- **Perception**: How an individual interprets the risks of a hazard.
- **Prediction**: The forecasting of the possible outcomes of a hazard event in terms of where, when, level of magnitude and potential impact on lives and property.
- **Risk**: The probability of a hazard event causing harmful consequences (expected losses in terms of deaths, injuries, property damage, economy and environment).

Disasters

You should be able to:

- distinguish between a hazard event and a disaster
- explain why this distinction is not always completely objective.

A hazard event is the occurrence of hazards (whether natural or human-made) that changes demographic, economic and/or environmental conditions. A disaster is an event that causes widespread disruption to a community beyond a point where it can cope without outside help.

You should be able to:

- describe the methods used to quantify the spatial extent and intensity of disasters
- explain the causes and impacts of any **one** disaster resulting from a natural hazard
- explain the causes and impacts of any **one** recent human-induced hazard event or disaster
- examine the ways in which the intensity and impacts of disasters vary in space and have changed over time.

Methods used to quantify the spatial extent and intensity of disasters include the Volcanic Explosivity Index (VEI), the Moment Magnitude Scale (MMS), although the Richter Scale is still widely referred to in the media, and the Saffir–Simpson Hurricane Scale (SSHS).

Impacts include loss of life, economic losses, insured losses and trauma. It is impossible to quantify all of the impacts of hazards.

Disasters appear to be becoming more frequent and their economic impacts greater (for example, the 2004 Asian tsunami and earthquake in Haiti in 2010). However, this may be the result of better recording of these hazards, or that, because of population growth, there are now more people at risk of hazards. The economic costs of hazards are rising, especially in richer countries where the value of properties is much greater than in poor countries.

Key definitions

- **Human-induced event**: An event resulting from human error, such as the Deepwater Horizon oil spill, the Union Carbide gas explosion at Bhopal and the Chernobyl explosion. Some of these events are pre-2000 but, as their impacts are still being felt today, they can be considered appropriate case studies.
- **Natural hazard**: An event occurring as a result of natural forces, for example, the Haitian earthquake in 2010.

Adjustments and responses to hazards and disasters

You should be able to:

- discuss the usefulness of assessing risk before deciding the strategies of adjustment and response to a hazard
- describe attempts that have been made to reduce vulnerability by spreading the risk (aid, insurance) and by land-use planning (zoning).

Risk analysis may help identify particular places and people for whom the risk of a hazard is greater. For example, maintaining a belt of mangroves reduces the impact of hurricanes on coastal communities. Areas built on unconsolidated rock are most at risk during an earthquake, such as in the San Francisco Bay area, especially in low-lying areas that have been reclaimed from the sea.

Be careful not to answer this question with reference to river flooding!

You should be able to:

- describe strategies designed to limit the damage from potential hazard events and disasters.

Strategies can include building design, buildings that are able to sway in tune with earth shocks, diverting lava flows, evacuating people (for example, in Florida and Louisiana during Hurricane Ivan in 2004), excluding people from an area (Montserrat's southern third has been an exclusion zone since 1997).

You should be able to:

- describe the range of responses, at the community, national and international levels, during and after a hazard event or disaster
- distinguish between rescue, rehabilitation and reconstruction responses
- explain how these responses are affected by individual and community perceptions
- examine the factors that affected the choice of adjustments before, and responses to, actual hazard events or disasters
- discuss the importance of re-assessing risk, and re-examining vulnerability, following any major hazard event or disaster.

Responses range from quiet acceptance of hazards to planning the management of the impacts of hazards. In some cases, people may leave an area because of the hazard (for example, Montserrat's population fell from around 12,000 to about 5,000 following the eruption of the Soufrière Hills volcano). Short-term responses are focused on search and rescue but this gives way to longer-term programmes of rehabilitation and reconstruction. These programmes can take years, as seen in the cases of Haiti and Myanmar.

Key definitions

- **Hazard assessment**: The process of determining when and where hazards have occurred in the past, their frequency and magnitude, and their likely impacts.
- **Long-term**: Factors or features of responses that occur much later, for example, redevelopment following an earthquake.
- **Medium-term**: Operations taking place between one week and a few months—providing accommodation, food and medical supplies are available.
- **Reconstruction**: This generally refers to a long-term rebuilding of the economy and society.
- **Rehabilitation**: An attempt to get people back into their homes.
- **Rescue**: The immediate response and attempt to get people to safety.
- **Risk assessment**: The process of establishing that a hazardous event of a particular magnitude will occur within a given period. Estimating the impact of the hazardous event, taking into account the location of buildings, facilities and emergency systems.
- **Short term**: Factors or features of responses that may happen quickly, for example, search and rescue in the first few days (less than one week) following an earthquake.

(a) Describe the difference between a hazard and a disaster. *[2+2 marks]*

(b) Explain why some sections of a community are more vulnerable to hazards than others. *[6 marks]*

(c) Compare the effectiveness of the methods used to predict the occurrence of two different natural hazard types. *[10 marks]*

[Taken from paper 2, May 2011]

How do I approach the question?

(a) "Describe the difference" means give a detailed account of the difference.

(b) "Sections of a community" means social groups, for example, rich and poor.

(c) The two natural hazards should be carefully selected to be sufficiently contrasting in their level of predictability in order to allow for meaningful comparison. Suitable examples would be more predictable hazards (hurricanes) and less, or unpredictable, hazards (earthquakes). Do not refer to both earthquakes and volcanoes in your answer as tectonic events are considered a hazard type.

Initially, you will need to describe at least two methods of prediction for each hazard before moving on to a comparison of their effectiveness. This means that you should cover their success and, possibly, their limitations. Although not a requirement of the question, reference to examples is essential if you are to achieve a high mark.

What are the key areas of the syllabus?

- Option D: Hazards and disasters—risk assessment and response
- Sub-topics: Characteristics of hazards; Risk and risk assessment

This answer achieved 18/20

This answer makes a clear distinction between a hazard and a disaster, but there needs to be some mention of outside help being necessary in the case of the latter.

(a) A hazard, natural or man-made, is any feature which poses a risk to the community or the environment near it. A disaster is a hazard event (the occurrence of a hazard) that causes widespread or unacceptable damage to the community, socially, economically, or environmentally. Hence we can conclude that a disaster is the realization of a hazard which creates devastation to the community or surrounding environment. [3]

(b) In any community, there are some sections which are more vulnerable to hazards than others. One of these secttions is those in the dependency ratio – that is, children and elderly – because of their fragility and the fact that they need outside help in case of emergency. Tourists or people from the outside (ex. New immigrants) are also more susceptible as they often lack the necessary information to protect themselves or evacuate. Furthermore, tourist dwellings are often built on areas with more risk, or lack the normal building regulations. Another especially vulnerable section of a community is the poor (or sometimes even destitute) first of all, especially in many less economically developed countries, those who are poor are often forced to live near or on hazardous regions because of the fact that they cannot afford more favourable circumstances. In some cases the benefits (such as more fertile land near a volcano) may lead these people to risk the potential; costs. Moreover, the lack of education and information would significantly lessen the preparedness of those with low incomes. Recovery after a potential hazard event is also difficult, for poor people do not have effective coping strategies or another form of back up (i.e. insurance), and hence will be left homeless and destitute. [5]

The vulnerable sections of a community are identified both by social group and physical area. Slightly more emphasis is needed at the local community level.

The student gives clear justification for the choice of hazards and covers predictive technology.

(c) The prediction of natural hazards is central to much debate and importance in the study of hazards by scientists. Two of these natural hazards are hurricanes and drought. For the former, much progress has been made in forecasting with technological advancement; the latter, on the other hand, has proven to be much more difficult to predict due to its duration and spatial extent.

The student explains the formation of hurricanes and refers to an example.

Hurricanes are natural hazards which are low-pressure systems characterized by high winds and heavy rainfall. Unlike hazards such as earthquakes and floods, the speed of onset of hurricanes is relatively slow-taking up to 14 days to form. This has made forewarning easier. The organization TIGGE has developed a system called the Global Interactive Forecast System, which attempts to track and forecast hurricanes. One of its earlier tests was utilized on the Typhoon Lupit. Satellite images taken of the typhoon showed it heading westwards towards the Philippines, though at a later time the tropical storm turned north east. Using forecast information, a strike probability map was produced and the potential places that could be hit issued warnings. The heavy thunderstorms forming Typhoon Lupit reached up to 15km, signifying heavy rainfall. Due to the early evacuation signals that were given, damage was limited to a minimum. As a rule, hurricanes are easy to predict as they form in certain conditions and certain time periods. Since warm temperatures are needed to allow their movements and wind currents, hurricanes form in oceans that has temperatures above 26°C, and only in certain latitudes near the Equator. Hurricanes tend to occur in autumn. All of these known characteristics of the hazards, in addition to satellite images and monitoring, have lead to relative success in hurricane prediction.

The origin and impact of drought are mentioned.

Drought is also a slow onset hazard; however, this fact often proves to be a disadvantage instead of benefit, for it has often been described as a "slow-creeping" phenomenon. This hazard occurs over a period of time, from a number of months to years, and often effects are not seen until it is well into Stage 2. Furthermore, drought does not have a spatial trend – it occurs in almost every climate regime, from Africa to the United States to Australia, and its impacts also vary greatly from region to region. However, there are some methods in which scientists have attempted to predict drought. One is by identifying teleconnections, or patterns which may induce drought, such as the El Nino Oscillation.

Strategies are linked to examples.

Some climates, such as the sub-Sahara Africa, are more susceptible to droughts. Since the first effects of drought are often those on the soil, systems are set up to closely monitor water table levels. In Australia, for example, government has implemented strategies in predicting regions that will be most affected by drought, such as measuring precipitation levels.

In general, of the two natural hazards, hurricanes are by far the easier to predict. This is due to the fact that while tropical storms formation has trends which are specific to its nature, the effects of drought are far-ranging and far-reaching. With the close monitoring of possible causes and plans to prevent these natural hazards, however, steps may be taken to improve drought prediction. [10]

Examiner report

Overall, this is an excellent effort and the student addresses all aspects of the question. The choice of the two types of hazard is justified and their characteristic features explained, with particular reference to their speed of onset. Predictive techniques are well explained and there is close reference to named locations throughout.

The conclusion is brief, but overall this is a very good and well-structured answer.

This answer achieved 13/20

The description of a disaster is very detailed and identifies clearly the severity of its impact on the need for outside help. However, there is no description of a hazard.

(a) There are multiple differences between hazards and disasters. Though disasters are after the consequences of natural and manmade disasters, vulnerability, the degree to which a hazard will effect a population due to their preparation, prediction and recovery, is an essential keyword addressing development and the consequences hazards will have for a certain population. A disaster occurs when a hazard has immense impact on the population, due to multiple events, happenings, that include the death of thousand of people, which could have possibly been prevented with better preparation and prediction, the destruction of infrastructure, the breakdown of vital and essential services, including health care and hospitals, which will limit and reduce the living quality of the affected in the long-term. The hazard itself is the event and hazards tend to have much greater impact on LEDCs that are clearly less prepared and need much longer, possibly many years to recover, and are therefore more vulnerable. Though disasters are caused by hazards the development of a country can distinguish and result in a possible disaster. MEDCs tend to be much less vulnerable to natural hazards as their infrastructure can often time resist the forces of the hazards, therefore less people die, less infrastructure is destroyed and the hazard will most of the time not result in a disaster, as it is often the case in LEDCs. [2]

Wealth is identified as a key factor controlling vulnerability, but reasons for vulnerability such as age and gender should have been given.

(b) Some sections of a community are much more vulnerable than others. This is due to the fact that poorer parts of the population are worsely affected by hazards. The infrastructure they live in is often times not resistant, while those who are wealthy enough do not directly suffer under the primary effects, i.e. collapsing of infrastructure, as the poorer population does. Those who tend to have less, the poor, have more to loose and are more vulnerable for multiple reasons. They live in lower quality homes, cannot afford sufficient food supplies after a hazard, are often times unaware of the hazard before it occurs and are less of an interest to assistance than the more wealthy. Since the poor are less prepared less informed about the hazard, they suffer more from the effects. The combination of these factors, poor infrastructure, high population density and little opportunity for self sustainment, clearly makes some sections of a community more vulnerable to hazards than others, who are wealthier. [3]

(c) When considering the effectiveness of two methods of predicting the occurrence of natural hazards such as earthquakes and hurricanes, one must distinguish between the general technology available for prediction in MEDCs and LEDCs. Furthermore, it is essential to keep and mind that predictions of hazards, whether it is earthquakes or hurricanes, are of no great importance when the population affected does not have access to this information and little knowledge of how to best prepare for such an event, especially if development is low and the hazards struck without warning.

Seismic activity is recorded before the earthquakes occur. Though in nearly all countries and areas suffering under high earthquake potential. This technology is available and used, seismic activity does not always imply a major earthquake is to follow. A certain time span can oftentimes be calculated for preparation to occur, but which time the earthquake will exactly strike are only assumptions. Plate boundaries that move and cause immense friction can suddenly release energy and and earthquake follows. Scientists can logically assume a moving plate boundary will result in a earthquake or possibly a volcanic eruption. The major problem with earthquakes is that the epicentre, the geographical point where seismic activity is highest, is not always situated on land, but also can be in the ocean resulting in tsunamis as it has recently been the case in Japan. Therefore making predictions as to what hazard will follow is very hard and [unprecise]. Additionally, when the populations is barely informed due to lack of telecommunications, infrastructure etc., these will be affected without warning and barely any preparation making past predictions with technology measuring seismic activity useless.

Hurricanes, however, can be more effectively predicted, as with the use of satellites their formation can be observed and recorded. Before making landfall the path the hurricane might take can be assumed and whether it is correct or incorrect the path can further be observed per satellite. Populations can be informed and preparation can occur. Though the impact of hurricanes are also oftentimes unexpected, with technology many assumptions about hurricanes having to do with their power and speed are oftentimes correctly calculated. This makes the methods used for predicting hurricanes more effective than for earthquakes.

When comparing earthquakes to hurricanes – it is obvious that both are highly powerful and can result in immense destruction of nature, infrastructure and the death of thousands of humans. Though earthquakes such as the Kobe earthquake in 1995 was correctly predicted and so was hurricane Nargis in Mynmar making landfall in 2008, the most important aspect to address is clearly vulnerability and the access to predicted hazard information in order to prepare, which tends to be highly insufficient in many LEDC, making methods and the effectiveness of prediction of hazards less relevant when considering vulnerability, as even with pre-knowledge about the future occurrence of hazards, LEDCs are most insufficiently prepared anyways and the hazard will strike and result in greater damage than in MEDC, cause an entire population simply cannot perfectly prepare for such an event, due to their low development and infrastructure. [8]

Examiner report

The student chooses two suitably contrasting natural hazards—earthquakes and hurricanes— and covers some aspects of their predictability, but more technical detail is required.

The emphasis of the argument is that lack of economic development can limit the ability of a country to apply predictive technology and adequately prepare populations. This is a thoughtful and well-argued answer.

11. Leisure, sport and tourism (option E)

This optional theme examines the characteristics and growth of the leisure industry, which includes a range of activities from informal recreation to more organized, commercial sport and tourism.

Leisure activities are examined in this optional theme on a range of scales from international to local. The growth of the leisure industry is both a consequence and a cause of economic development.

Improvements in socio-economic conditions, education and transport have allowed the increasing participation of populations from poorer countries in international sporting activities. Nevertheless, poverty still excludes some countries benefiting from the prestige and status gained through participating and hosting international sporting events.

Improved transport, communications and economic progress in low- and middle-income countries have made them increasingly attractive as tourist destinations. The benefits may be significant, and some economies have allowed tourism to take over as the primary economic activity. Managed sustainably, tourism can be an important stimulus to economic development, but ill-managed mass tourism can have serious adverse consequences for the host countries. In this optional theme, case studies are chosen to illustrate the successes and failures of tourism development and to establish ways in which the industry may be sustainably managed. Examples are drawn from a range of geographical locations including small island economies, such as Mauritius, to areas such as Spain, where tourism has a long tradition but for whom competition from more exotic locations is now a threat. Management issues are concerned with protecting the environment from damage, minimizing economic leakage from the host country and avoiding cultural dilution. These concerns are embodied in the concept of sustainable tourism management, which is now the aim of many developing tourist destinations.

Leisure

You should be able to:

- discuss the difficulties in attempting to define leisure, recreation, tourism and sport
- discuss the influence of accessibility, changes in technology and affluence upon the growth of these activities.

You will need to learn the definitions of each of the four types of leisure and be able to distinguish differences between each of them. Sometimes people can participate in sport, recreation and tourism at the same time. For example, tourists may be involved in playing golf while on holiday, or a participant in the Olympic Games may visit a health spa.

Key definitions

- **Leisure**: Any freely chosen activity or experience that takes place in non-work time.
- **Recreation**: A leisure-time activity undertaken voluntarily and for enjoyment. It includes individual pursuits, organized outings and events, and non-paid (non-professional) sports.
- **Sport**: A physical activity involving a set of rules or customs. The activity may be competitive.
- **Tourism**: Travel away from home for at least one night for the purpose of leisure. Note that this definition excludes day-trippers. There are many possible subdivisions of tourism. Sub-groups include:
  - eco-tourism—tourism focusing on the natural environment and local communities
  - heritage tourism—tourism based on a historic legacy (landscape feature, historic building or event) as its major attraction
  - sustainable tourism—tourism that conserves primary tourist resources and supports the livelihoods and culture of local people.

Leisure at the international scale: tourism

You should be able to:

- explain the long- and short-term trends and patterns in international tourism.

This means that you should know the historical trends of global tourism and the rates of change for the different world regions.

You should be able to:

- examine the changes in location and development of different tourist activities
- explain the growth of more remote tourist destinations.

You should understand the differences between mass tourism and activities that are less damaging, such as ecotourism.

Note that tourist destinations can be on any scale, from continental to local. Your explanation should relate to the following factors.

- Growing demand from foreign and domestic tourism.
- Improved supply of tourist resources, such as infrastructure, accommodation and leisure activities at the destination.
- Improved accessibility through expansion of the transport network.

Key definitions

- **Domestic tourism**: Tourism at home involving residents of the country.
- **International tourist arrivals (inbound tourists)**: Tourists arriving from another country.
- **Primary tourist/recreational resources**: The pre-existing attractions for tourism or recreation (that is, those not built specifically for the purpose), including climate, scenery, wildlife, indigenous people, cultural and heritage sites. These are distinguished from **secondary tourist/ recreational resources**, which include accommodation, catering, entertainment and shopping.
- **Secondary tourist/recreational resources**: Human-made tourist attractions in the area, such as holiday camps, leisure centres and golf courses.

Leisure at the international scale: sport

You should be able to:

- examine the social, cultural, economic and political factors affecting participation and success in two major international sports.

"Social" refers to custom, religion, gender and age factors; "cultural" refers to ethnicity; "economic" refers to the level of wealth; political factors may result in banning some participants. For example, cricket is most popular in former British colonies, and the high altitude of parts of Ethiopa produce long-distance runners with higher concentrations of red blood cells, which may give them an advantage. You should refer to at least one example to explain each factor.

You should be able to:

- analyse the geographic factors that influenced the choice of venue
- examine the factors affecting the sphere of influence for participants and supporters
- evaluate the short- and long-term geographic costs and benefits of hosting such an event at both the local and national level.

These points relate to a case study of a contemporary international sports event.

Geographic factors favouring the location of venue include low site costs, physical suitability, access to natural and human resources and political security. "Analysis" involves further breakdown of the factors and explaining the relative importance of each. "Examine" is a higher-order command term, which means that you will need to look at all types of factors relevant to an international sporting event. The "sphere of influence" is the geographical area from which participants and supporters are drawn; the more important the event, the larger its sphere of influence. There is plenty of scope for a well-developed answer about the benefits of hosting a major sporting event, but the final exam question may be more open-ended and offer fewer prompts. Choose a recent event, but one that provides sufficient data for you to evaluate both short- and long-term impacts.

Leisure at the national/regional scale: tourism

You should be able to:

- examine the economic, social and environmental impacts of tourism.

This relates to a case study of a national tourist industry. The following impacts should be examined.

- Economic benefits—these include tourist revenue and the multiplier effect.
- Economic costs—these include leakage of revenue through use of non-national transport and imported goods for tourists.
- Social benefits—these include cultural integration, improved language and technicalskills and improved infrastructure.
- Social costs—–these include cultural dilution, moral degradation, and disease.
- Environmental benefits—these include investment, and conservation of natural habitats and wildlife.
- Environmental costs—these include aircraft emissions, waste overload, water shortage, landscape damage, loss of natural habitats and reduction in biodiversity.

You should be able to:

- evaluate the strategies designed to manage and sustain the tourist industry.

This relates to a case study of ecotourism. A definition of ecotourism is essential and it must involve sustainable practices affecting both the environment and local people. Knowledge of a suitable ecotourism case study is important here. You should describe, explain and evaluate more than one strategy designed to promote sustainability and improve national prestige and recognition.

You should be able to:

- examine the importance of tourism as a development strategy for low-income countries.

Tourism allows many low-income countries to develop economically, especially where they have a rich cultural heritage or attractive physical features. Many small island economies, such as the Maldives, have become dependent on tourism and are vulnerable to falling tourist demand resulting from shocks such as natural disasters, international terrorism, global pandemics and financial crises.

Leisure at the national/regional scale: sport

You should be able to:

- explain the hierarchy of a league and the location of the teams
- examine the relationship between team location and the residence of its supporters.

A hierarchy is a ranked arrangement of the participants or teams in a sport, for example, the UEFA Europa League.

The study of a national sports league is fundamental to this sub-topic.

Leisure at the local scale: tourism

You should be able to:

- For **one** named city or large town:
  - describe the distribution and location of primary and secondary tourist resources
  - discuss the strategies designed to manage tourist demands, maximize capacity and minimize conflicts between local residents and visitors, and avoid environmental damage.

You may be asked to draw and annotate a sketch map of your chosen city or town to show the location of primary resources and secondary resources (monuments, museums, cathedrals and other heritage attractions) or, alternatively, interpret a given map extract.

Crowd control near tourist hotspots can be achieved by restricting access to tourist sites in time (with limited opening times) or in space (through zoning). Both can be socially, economically and environmentally successful. You will need to refer in detail to case studies.

You should be able to:

- examine the concept of carrying capacities in a rural tourist area
- discuss strategies designed to maximize capacity and minimize conflicts between local residents and visitors, and avoid environmental damage.

Carrying capacity is the maximum number of visitors/participants that an event/site can satisfy at one time. It is a dynamic concept that can change both in time and space. You will need to refer in detail to case studies.

Leisure at the local scale: sport and recreation

You should be able to:

- explain the relationship between urban settlements and recreational and sports facilities in terms of frequency, size, range and catchment area.

Large, wealthy cities have a wide range of sporting and recreational facilities. Some higher-order facilities, such as major sports stadia, draw upon a large catchment area, and as their spheres of influence may be national or even international, such facilities may include an Olympic-size swimming pool. Further down the leisure hierarchy are small towns offering lower-order facilities and serving local residents.

You should be able to:

- examine the distribution and location of recreational and sports facilities in urban areas and relate the patterns to accessibility, land value and the physical and socio-economic characteristics of each urban zone (from the central business district to the rural–urban fringe).

An annotated sketch map showing recreation and sports facilities relative to the central business district (CBD) and the suburbs could be helpful to support your argument. Accessibility, land value and the characteristics of each urban zone determine where sports or recreational facilities should be located.

You should be able to:

- discuss the role of sport and recreation in regeneration strategies of urban areas.

There are many examples of regenerative transformation taking place, mostly in the low-income urban districts characterized by dereliction, decay amd social deprivation. The discussion should consider the costs and benefits of projects. Case studies are essential.

Sustainable tourism

You should be able to:

- define sustainable tourism
- examine the extent to which it might be successfully implemented in different environments.

Affluent countries with well-educated populations are more likely to have well-established programmes promoting sustainable tourism. However, extreme environments with poor populations and fragile ecosystems, such as those found in the African arid and semi-arid zones, are less likely to be able to adopt sustainable practices until tourism is well established.

Key definitions

- **Sustainable tourism**: Tourism that conserves primary tourist resources and supports the livelihoods and culture of local people.
- **Tourism**: Travel away from home for at least one night for the purpose of leisure. Note that this definition excludes day-trippers. There are many possible subdivisions of tourism. Sub-groups include:
  - eco-tourism—tourism focusing on the natural environment and local communities
  - heritage tourism—tourism based on a historic legacy (landscape feature, historic building or event) as its major attraction
  - sustainable tourism—tourism that conserves primary tourist resources and supports the livelihoods and culture of local people.

(a) Describe the main characteristics of sustainable tourism. *[4 marks]*

(b) Explain **three** different impacts of tourist developments on the environment. *[3×2 marks]*

(c) Examine how tourism has had social and economic impacts on **one** country. *[10 marks]*

[Taken from paper 2, May 2011]

How do I approach the question?

(a) The key characteristics of sustainable tourism are preserving the environment and the livelihoods of local communities.

(b) You should choose three distinct environmental impacts. Examples include impacts on water, land, atmosphere and ecosystems.

(c) There are several key words in the question that deserve special attention. "Examine" means explain with reasons and justify your views. "Social impacts" relate to people at a tourist destination, and the effect tourism has on their quality of life and culture. "Economic impacts" involve money and the income of a host population. Remember to give this question the widest possible interpretation. For example, impacts may be local, regional or national. They may also be long or short term, positive or negative. The question asks for several impacts, but only in one country. You are, therefore, expected to have extensive factual knowledge of a case study here.

What are the key areas of the syllabus?

- Option E: Leisure, sport and tourism
- Sub-topics: Leisure at the national/regional scale; tourism; Sustainable tourism

This answer achieved 18/20

A good answer that recognizes the essential features of sustainable tourism: careful resource conservation and waste management, minimal environmental impact and maximum involvement of local communities.

(a) Main characteristics of sustainable tourism include, more sustainable use of resources to ensure long term environmental and economical benefit to reduce waste and over consumption to reduce waste from local communities and provide environmental sustainability. to increase involvement of local communities and economies to ensure long term sucess of industry while preserving local cooperation in environmental and social sustainability. they also take into account the concept of carrying capacity which is divided into three where all ensure the environment quality of the area while pleasing tourist and peasants alike to increase the learning experience of both. [4]

The impacts upon water supply are well explained with brief reference to Spain and Zuni Pueblo (not located). There are further useful examples on environmental impacts covering vegetation loss in the Alpine region. The consequences of vegetation loss are well explained.

(b) It affects availability of water and the consistant uptake of water in area such as Spain and Zuni pueblo tourism sites can lead to failing quality of soil and its ability to sustain communities leading to biodiversity loss. This is why in Zuni pueblo tourism is considered to some extent dangerous, leading to more control by locals while in Spain the exact opposite has been happening, leading to desertification vegetation is affected by tourism, especially in Alpine periglacial environment where tourism development infrastructure and social development plans have lead to vegetative loss, which in turn leads to the melting of the perma frost due to decreasing albedo affect and exposure to extreme climate. Deforestation to build secondary resource for tourism attraction such as hotels and accommodation facilities have led to loss in bio diversity and suseptable to more surface erosion. This is present in the Omu regions of Ethiopia, where growing tourism has led to more deforestation which in turn led to loss of bio diversity and less water infiltration by locals. [6]

(c) Even though Tourism has and still been in the growing stages in Ethiopia, tourism has had both social and economic properties. The exploration days of the butlers model of tourism evolution have ended and involvement by local government old communities have been key in the growing success of Ethiopia as an area to visit. However the implications that come along with tourism, both the negative and positive has been noticed.

Historical sites such as Axum and Lalibela have attracted Historians from predominantly Europe and America for over 20 years and this has led to the local involvement of farmers and traders in the attractive industry. It has provided them a source of income in an area where drought and famine is common. More developed needs and accommodation areas have been developed and therefore it is in the involvement stage of Butlers model. However local resentment is present to some extent as a result of agriculture land being taken and homes being destroyed to build roads. Therefore it has brought economic advantages but has resulted in social detest. Therefore there is a more urgent call by locals to step back and asses the social implications of a growing tourism business in these areas. However growing development schemes are underway, schemes that also provide job opportunity, an argument that is mostly used by local developers.

Natural tourism in Ethiopia has taken two different routes. In the lake tana region, the lake that feeds the Nile, tourism has been growing around what is known as Tis Abay, or the smoky waterfall. This beautiful site in a secluded, green and mostly very dispersed farming settlements around it has made tourism development green and sustainable. One could argue its because of it being in exploration stage, but local farmers involvement and growing tourist number says otherwise. However in the one region in the South, environmental tourism sites have been problematic for locals who can not provide security for there cattles against predetaros in the national park nearby. The growing economic benefit doesn't reach these people and therefore they have been affected negatively in both political and economic sense, creating resentment and distrust with local officials and governmental development schemes.

Therefore, in Ethiopia, the economic impact of tourism has been beneficiary for the country as a whole but not for the locals who loose their homes and livehood good due to this business. Therefore tourism has had a positive increase for country is economy while social impacts of tourism are negative and is now being tried to be controlled by local residents. [8]

Examiner report

Different impacts, both positive and negative, are recognized and identified in particular areas of Ethiopia covering benefits for farmers. The focus is on the economic side and less on social impacts. The answer is informative and fulfills the requirements of the question well.

This answer achieved 9/20

A non-geographical term that needs explanation.

The student touches upon economic and social characteristics, but description is inadequate.

(a) The main characteristics of sustainable tourism are: Environmentally Friendly, Attractions that have a constant demand, which means constant tourist attraction, Avoiding Money Drainage due to foreign companies that cause the citizens to lose their jobs, and the money earned leaves the country, and Cultural Integration. [2]

(b) Environmental Damage:

Due to littering national parks, destroying coral Reefs. And the Destruction of the environment in order to create the infrastructure for more tourist attractions.

Three acceptable impacts, but no explanation.

Environmental Development:

Due to creating National Parks for Tourists to visit, creating Exclusive Economic Zones to increase their Beauty of the land, in order to attract more tourists. [3]

Dubai is not a country it is a city in the United Arab Emirates.

(c) The economic impacts of tourism on Dubai were immense. They have seen a massive increase in economic development, due to their high rate of creating new infrastructure for more and more tourist attractions. Since Dubai is located in the desert, and their main income comes from Oil, switching their economy to tourism was the best they could have done, since the Oil is going to run out soon, and if that happens they will still have the tourism on which they can rely on. Unlike in other countries like Tanzania where TNC's earn most of the money, and therefore money draining out of the country, most of the money earned in Dubai stays there, therefore the economic stability of Dubai is steadily increasing. However the social impacts of this are very different. The Westernisation causes a lot of Cultural Diffusion, when a lot of the Arabic traditions are getting lost. Since tourists wear bathing suits while swimming, they are not following Arabic tradition, which makes Muslim women feel discriminated since they do not have the same rights. Therefore tourism might cause negative social impacts. [4]

Examiner report

The choice of Dubai is inappropriate and, although some valid references are made to economic and social impacts, they are generalized. There is no conclusion.

12. The geography of food and health (option F)

This optional theme considers the issue of health and its relationship with food supply and disease. International variations in health are observed using data such as calorie intake or infant mortality rate. Success in combating disease is measured through the United Nations' Millennium Development Goals.

Patterns of global food availability and the imbalance between supply and demand are recognized and the effect of food insecurity evaluated. New agricultural technology and its effects are examined through food output and environmental impact.

The study of disease requires an understanding of the nutritional and epidemiological transition that countries make as they progress economically. Changes in lifestyle and improvements in the standard of living are desirable, but inevitably increase resource consumption. These changes may induce environmental pollution, a highly calorific diet and increased incidence of chronic disease such as diabetes, lung cancer and cardiovascular disease. Therefore, the burden of disease can weigh heavily on these economically emerging countries.

Detailed case studies are essential, especially when impacts and evaluation are required. You must know case studies of **two** diseases chosen from two **different** categories out of the following three: vector-borne, water-borne or sexually transmitted disease. Certain diseases, such as malaria, can fulfill several of the optional theme's requirements, but detailed knowledge is essential.

Health

You should be able to:

- describe the variations in health as reflected by changes in life expectancy at national and global scales since 1950
- explain the patterns and trends in terms of differences in income and lifestyle.

It is important that you can describe relationships between variables by comparing their global pattern on a map, analysing trends on a line or bar graph and describing the strength of their relationship by interpreting a scatter graph or correlation coefficient such as the Spearman rank.

You should be able to:

- evaluate life expectancy, infant mortality rate (IMR) and child mortality, HALE (health-adjusted life expectancy), calorie intake, access to safe water and access to health services as indicators of health.

The command term "evaluate" means you should explain how well each of these measures indicate the well-being of a specific population. You should learn statistics for two countries at contrasting levels of development for each of the indicators listed.

You should be able to:

- discuss the geographic factors that determine the relative emphasis placed by policy-makers, in **one** country or region, on prevention as opposed to treatment of disease.

This is a complex issue and you will need to discuss the geographic factors that determine how governments manage disease in a country, such as income availability, the extent of the disease, its pattern and speed of diffusion, the public image of the country, its level of global participation and the impact of the disease on the economy.

Key definitions

- **Access to health services**: This can involve several measures, such as the number of people for each doctor, health worker or hospital bed.
- **Access to safe water**: Access to clean water that is affordable, in sufficient quantity and is available without excessive effort and time.
- **Epidemiological transition**: This describes the decline in infectious disease and the rise in non-communicable disease that may occur in a country as it becomes more developed.
- **HALE**: Health-adjusted life expectancy, based on life expectancy at birth but including an adjustment for time spent in poor health (due to disease and/or injury). It is the equivalent number of years in full health that a newborn can expect to live, based on current rates of ill health and mortality.
- **Incidence**: The number of cases of a disease that occurs in a country or area over a period, usually a year.
- **Infant mortality rate (IMR)**

$$= \frac{\text{total no. of deaths of children} < 1 \text{ year old}}{\text{total no. of live births}} \times 1000 \text{ per year}$$

- **Infectious or communicable diseases**: Also known as diseases of poverty, these may be contagious and are transmitted through close human contact or by vectors. Examples include cholera and malaria.
- **Life expectancy**: The average number of years that a person can be expected to live, given that demographic factors remain unchanged.
- **Non-communicable, chronic diseases**: These are non-infectious, long-term and degenerative diseases that are usually associated with old age, but they may **be induced by lifestyle** and can affect younger people too. The most common diseases are heart disease, stroke, chronic respiratory failure and cancer.
- **Prevalence**: The percentage of a country or area's population suffering from a disease.

Food

You should be able to:

- identify global patterns of calorie intake as one measure of food availability
- distinguish between malnutrition, temporary hunger, chronic hunger and famine
- discuss the concept of food security.

You should be able to define calorie intake and recognize global patterns of consumption on a map.

The terms relating to hunger are defined under "Key definitions" below and have distinct differences. Malnutrition and hunger may be referred to as chronic, which means that they are long-term conditions. However, famine and hunger may also be temporary and acute. Make sure you are clear about the meaning and use of all these terms.

The United Nations' Food and Agriculture Organization (FAO) has the following definition of food security:

"Food security exists when all people, at all times, have access to sufficient, safe and nutritious food to meet their dietary needs and food preferences for an active and healthy life."

A discussion of food security should recognize that it's not simply a matter of calorie intake and food quantity—quality is equally important.

You should be able to:

- explain how changes in agricultural systems, scientific and technological innovations, the expansion of the area under agriculture and the growth of agribusiness have increased the availability of food in some areas, starting with the Green Revolution and continuing since
- examine the environmental, demographic, political, social and economic factors that have caused areas of food deficiency and food insecurity.

Scientific change means using biotechnology and high-yielding varieties (HYVs). Technological innovations include mechanization, fertilizers, pesticides, drainage, irrigation and expansion of the farmed area.

The factors causing areas of food deficiency and food insecurity are complex.

The term "food deficiency" means inadequate supply, and this may result from poor yields owing to drought, flood, conflict or inefficient farming practices. Food insecurity occurs because a population cannot get hold of a reliable food supply. Low income or political barriers may prevent people accessing food even though it is available. This means that famine is not always associated with poor food production.

Food production may increase in a country, but per-capita consumption may decrease owing to a relatively fast population growth rate.

Within one country there can be great variation in the pattern and level of food insecurity, which is often dependent on income and availability of transport networks.

You should be able to:

- examine the variety of causes responsible for a recent famine.

A national case study of one country is essential here. The case study must be recent (not before 2000).

You should be able to:

- examine the impacts at a variety of scales of trade barriers, agricultural subsidies, bilateral and multilateral agreements, and transnational corporations (TNCs) on the production and availability of food.

You should explain all these trade-related terms and the development of surpluses and deficits in food production that may result. These trading policies may be implemented at a national or international level, but they can impact upon the individual farmer and a local area.

The effect of agribusiness and the influence of TNCs on food production in **one** African country should be learned.

You should be able to:

- evaluate the relative importance of food aid, free trade and fair trade in alleviating food shortages.

You should compare the extent to which each of these three approaches help to avert or alleviate food shortages. Do not simply describe and explain. You must rate these approaches according to their effectiveness.

You should be able to:

- examine the concept of sustainable agriculture in terms of energy efficiency ratios and sustainable yields
- examine the concept of food miles as an indicator of environmental impact.

Many modern industrialized agricultural systems have high energy inputs relative to their outputs, so such systems are energy inefficient and non-sustainable. You will need to give examples of energy ratios for a variety of agricultural systems. Sustainability is achieved through conserving natural resources and minimizing pollution in order to maintain food yields for the future.

You should define the concept of food miles, then give examples of different foods and how their place of origin, mode of transfer and distance to their destination affect their food miles rating.

Key definitions

- **Agribusiness**: Large-scale industrialized agricultural production. Agribusiness controls planting, breeding, harvesting, slaughtering, processing, packaging, transport and marketing. It often involves transnational corporations (TNCs).
- **Fair trade**: Also known as ethical trade, this is trade that attempts to be socially, economic and environmentally responsible.
- **Famine**: An extreme shortage of food resulting in mortality. Famine can occur where food is available but where people do not have sufficient money or exchange entitlements to buy it.
- **Food aid**: This is given in emergencies to countries with an acute food shortage, or as long-term assistance that can alleviate shortage but depress market prices for local farmers.
- **Food miles**: A measure of the distance that food travels from its source to the consumer. This can be given either in units of actual distance or of energy consumed during transport.
- **Food security**: Food security exists when all people, at all times, have access to sufficient, safe and nutritious food to meet their dietary needs and food preferences for an active and healthy life. (Source: FAO)
- **Free trade**: This allows a country to trade competitively with another country. There are no restrictions regarding what can be exported or imported. By contrast, **protectionism** creates restrictions to trade on imports as well as exports.
- **Hunger**: A shortage of food usually resulting from inadequate supply and poverty. It may be a temporary or chronic condition. In the latter case, it can eventually lead to starvation and death.
- **Malnutrition**: This results from a diet with an imbalance of proteins, energy and minerals. Mineral deficiency may lead to diseases such as kwashiorkor and calorie/energy deficiency to marasmus. Calorie/energy excess may result in **obesity**, an unhealthy condition where excess body fat has accumulated and the body mass index (BMI) exceeds 30 kg/m^2.

Disease

You should be able to:

- explain the global distribution of diseases of affluence
- explain the global distribution of diseases of poverty.

You should describe and explain the global pattern of **two** chronic, non-communicable diseases resulting from the lifestyle changes associated with economic development. Good examples include cardiovascular disease, lung cancer and diabetes.

You should describe and explain the global pattern of **two** infectious diseases associated with poor living conditions, such as a lack of fresh water, sanitation, overcrowding and inadequate food supply. Good examples include measles, kwashiorkor, schistosomiasis and cholera.

You should be able to:

- explain how the geographic concepts of diffusion by relocation and by expansion apply to the spread of diseases
- examine the application of the concept of barriers in attempts to limit the spread of diseases
- describe the factors that have enabled reduction in incidence of a disease.

There are several well-documented case studies that may illustrate aspects of diffusion, including SARS in 2003 and H1N1 (swine flu) in 2009.

Barriers may take several forms, from physical to behavioural. For example, mosquito nets are an effective physical barrier to malaria, whereas vaccination is an effective barrier against the spread of measles. Behavioural barriers to the spread of disease can be encouraged by health education; anti-smoking and safe sex campaigns are good examples.

You should consider the economic, environmental, sociological and political factors that have helped to reduce the number of cases occurring over time of **one** disease. Reliable case studies include cholera, smallpox, polio, leprosy and typhoid.

You should be able to:

- examine the geographic factors responsible for the incidence and spread of **two** diseases
- evaluate the geographic impact of these two diseases at the local, national and international scales
- evaluate the management strategies that have been applied in any one country or region for one of these diseases.

Choose any **two** diseases, each one from a different category. Good examples include malaria, which is vector-borne, cholera, which is water-borne and HIV/AIDS which is sexually transmitted. Some of the geographic factors responsible for their incidence and spread, such as poverty and population mobility, are common to all, whereas malaria and cholera are also influenced by environmental factors. Cultural factors are important with HIV/AIDS.

Geographic impacts may include population mortality, fertility, internal migration and economic productivity at the local and national scales, and economic participation, migration and tourism at the international scale. Remember to mention that the impacts are usually negative, but may be short or long term.

You should evaluate the management strategies for one of your chosen diseases.

The graph shows the rates of disease in a city in India between 1991 and 2001.

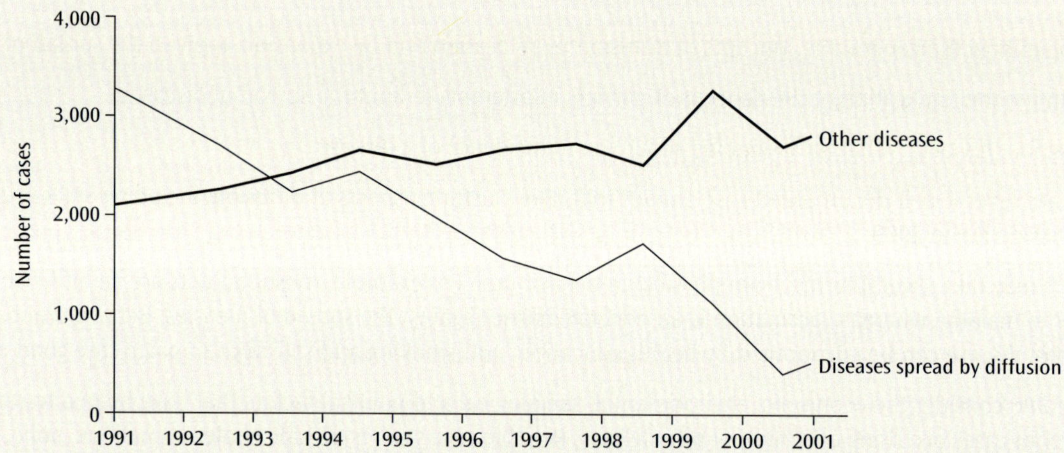

Source: Jayasree De. (2007). This Changing World: Development, Environment and Urban Health in India. *Geography*, volume 92, issue 2, pages 158–160.

(a) (i) Describe the trend in diseases spread by diffusion between 1991 and 2001. *[2 marks]*

(ii) Describe what is meant by "diffusion by relocation" with reference to a disease. *[2 marks]*

(b) Explain the global distribution of diseases of poverty. *[6 marks]*

(c) Referring to **one or more** diseases, discuss the factors that determine the relative importance of policies of disease prevention as opposed to policies of treatment. *[10 marks]*

[Taken from paper 2, May 2011]

How do I approach the question?

(a) Use all the data provided on the graph. That means looking at the trend of diseases spread by diffusion and relating it to other diseases.

(b) Although it does not ask you to describe, it is an essential introduction to your explanation.

(c) You are asked to "discuss the factors that determine the relative importance". This means looking at the factors that determine medical policies and explaining why treatment is more commonly adopted than prevention (and vice versa) with some diseases. Remember that the focus is on **factors** throughout. These might include the seriousness of the disease, its national prevalence and incidence, its diffusion rate, the costs of treatment and prevention, and the national spending priorities of the government.

You should choose your disease(s) carefully so that it (they) fit the requirements of the question. One disease may be diabetes, which can lead to serious cardiovascular complications that are costly to treat. In the case of poor countries, preventive policies, such as health education, may be more effective and offer the only practical solution. The policy is adopted according to the national funds available and, in some countries, both prevention and treatment policies are applied. Named examples of one or more countries are expected in the answer.

What are the key areas of the syllabus?

- Option F: The geography of food and health
- Sub-topic: Disease

This answer achieved 18/20

The student makes two valid observations by looking at trends and making a comparison.

(a) (i) The trend seen in diseases spread by diffusion between 1991 and 2001 has been an overall decrease from over 3000 cases in 1991 to less than a 1000 in 2001. The diseases spread by diffusion have also dropped below the incidence of other disease in that city. [2]

A competent answer that shows a very good understanding of disease diffusion.

(ii) Diffusion by relocation refers to a type of spreading of a disease, where the disease leaves a region of origin into another region. Such a case would be that of an HIV infected person who moves from one country to another. [2]

An excellent answer that identifies the link between the spread of communicable disease and poor levels of hygiene, which ultimately threatens survival.

This student's answer goes well beyond the expectations of the question. The student may have been awarded full marks for writing considerably less than this.

(b) Disease of poverty are diseases which can be caused and treated with medical attention but still occur in regions of the world where proper medical attention is too expensive for the local population or non-existent due to a lack of technology and infrastructure.

Many diseases of malnutrition are diseases of poverty and in most cases in Sub-Saharan Africa and South-East Asia. Disease such as Marasmus and Kawasaki can often be treated and cured from those suffering with a better diet, balanced carbohydrates, proteins, lack of vitamins and minerals. However those afflicted are those in poor regions of the world where subsistence farming is the only means for survival and there is very little diversity in the food being consumed, these diseases are also often found in regions where culturally only one food is mostly eaten, and as in South-East Asia where most of the diet is carbohydrate but lacking in proteins and fats, resulting in many vitamin deficiencies.

Water-borne disease such as cholera are also found to be diseases of poverty, where unsafe drinking water is the source of illness. These are yet again founds in the poorer regions of the world mostly Sub-Saharan Africa, as many people must drink muddy, stagnant waters to survive, whilst richer countries have had the time and economic power to provide safe water. Furthermore, there are also insect-spread diseases such as Malaria, which affect the poorer regions of the world because they have not been able to rid themselves of Malaria infected mosquitoes. Europe and Northern America rid themselves of malaria using insecticides designed to kill these mosquitoes, have also developed treatments which has cleansed most of these rich countries from Malaria. Whilst poorer countries, in regions where Malaria mosquitoes thrive, still suffer millions of deaths because the populations cannot afford treatment and the government is not able to take large scale action to rid themselves of their poverty caused disease. [6]

(c) Policies regarding diseases are often put in place with regards to many factors such as the incidence of a disease, its danger whether or not there exists treatments. As such policies of prevention often vary greatly from policies of treatment concerning many diseases.

HIV/AIDS is a disease where the policies of prevention vary greatly from those of treatments. This is due to the virulence and danger which HIV poses to the global population and also the fact that there is not yet a definitive cure for the virus, only retro-virals which slow down the effects of the HIV virus. As such policies concerning the prevention of HIV have gotten most attention, the International AIDs Organisation focus much more on spreading information on how to protect oneself from contracting HIV rather than how live with. There is ample information on the precautions one must take when there is the possibility of mixing bodily fluids. Countries have also opted for a greater use of preventional policies. The United States until recently banned the entry of HIV infected people into the country to try and prevent the spread of the virus within the United States. Countries in Europe such as Switzerland also has very good treatment plans, with modern therapies using retro-virals available to help infected patients lead longer and better lives. As such HIV would be a disease where policies are focussed on prevention due to the factors that there is no cure, incidence and prevalence are high, thus prevention is the better option as a widespread policy. Prevention information also breaks the stigmas concerning HIV/AIDS as people know how to protect themselves and are no longer fearing irrational things.

The focus is limited to rich countries.

Diabetes is a disease which has a mixed number of policies which focus both on prevention of type II diabetes caused mostly by an unhealthy diet and type I diabetes which is usually a genetic predisposition. Diabetes will have policies of prevention, where a healthy diet and regular exercise are promoted, and doctors advise all potential patients that type II diabetes can be prevented with a healthy lifestyle. Diabetes also has policies for treatment which are very effective, the development of insulin injections has saved millions of lives and all richer countries have available treatments so that those with diabetes can still keep on living normal lives. The reasons for the balance between prevention and treatment is due to the fact that treatments are readily available, yet in many cases the disease can be avoided. As such we find that most countries both try to prevent Diabetes but are also able to provide proper treatment to their populations. [8]

Examiner report

The student identifies several factors that control the approach to treatment, and uses HIV/AIDS and diabetes as examples. These are very well chosen and illustrate the importance of a number of factors, such as the extent to which the disease can be cured, its incidence and prevalence. This answer is informative and relevant. However, the answer was not awarded full marks because there was no conclusion, which might have reflected upon the economic status of the country as a major factor controlling policies of treatment.

This answer achieved 10/20

The student recognizes that the number of cases of diseases spread by diffusion declined between 1991 and 2001 and relates this to other diseases.

(a) (i) The graph shows the trend in disease spread by diffusion between 1991 and 2001 in India. By looking at the graph one can see that the number of cases and disease spread by diffusion has declined drastically from 3200 causes in 1991 to about 500 cases in 2001. However the number of causes of other disease (which can come from contaminated water) increased slowly in 1991 from 2050 cases to about 2500 cases between 1998 and 1999. After that half a year later there was a drastic increase to about 3200 cases which however decreased to around 2700 cases in 2001. [2]

The correct definition is given.

(ii) Diffusion by relocation is referred to the spread of disease by a change in location. For example AIDS has been spread worldwide because of people moving to different places and infecting more people. [2]

The relationship between disease and poverty is clearly stated, with a focus on Sub-Saharan Africa, but there is no overview of global distribution. The reasons given are limited to poor education (for example, AIDS) and the low status of women. The answer shows only a limited understanding of factors affecting the global pattern of diseases of poverty.

(b) The global distribution of disease of poverty is mainly seen in less economically developed countries for example countries in the sub-Saharan Africa. This is due that countries in the sub-Saharan Africa are known to be the poorest where an average income is less than a dollar a day. This directly indicates an lack of education. For example in the case of AIDs. Botswana has 36% of its population affected with AIDs and 60% of children below the age of 5 affected by AIDS. This mainly due to the lack of contraseptions, sex education, and the role of women. Especially in rural areas the empowerment of women is very low and men are usually with many different women which increase the risk of HIV/AIDs infections. Even though more economically developed countries such as the USA are also affected by AIDs rates are not as high as in country of poverty mainly because of education. [2]

(c) AIDs is one of the most common and widespread disease in the sub-Saharan Africa. There are many factors that contribute to the prevention of AIDs. Botwana has one of the highest rates of AIDs and a very low life expectancy of 42 years. If precautions and prevention methods are taken in to consideration the country will have a possibility to develop. There are many methods to hinder the infection of AIDs. The most important factor is Education. Botswana had a very low literacy rate with the dominant number of literal people being men. If education is granted sex education will also be taken into consideration. People, men and women will be better informed about contraception and condom use. This is the major first step to improve health and rates of people infected with AIDs will decrease. Women empowerment, or rather the lack of women empowerment and gender inequality must also be changed. This requires a change in culture, however, women are opposed to polygamy which increases the rate of women getting infected by the male or the male being infected by other women. Like this disease spreads rapidly, therefore if women gain more power and more rights situation like these can be hindered. Some young Africans generally tend to think of AIDS as a minority as people dies from other infectious diseases. However the way they think must be changed as the better your immune system the more chance your body has to combat other sicknesses or diseases. If factors like I mentioned above improve and change it will also allow Botswana to develop. There will be a higher life expectancy, a higher literacy rate for both male and females, which will allow women's rights to be reinforced and more jobs and employment and life opportunities will be granted to the locals. [4]

Examiner report

This answer is limited and does not address the question fully. It needs to focus on the **relative** importance of the two policies. This means assessing the nature and costs of treatment as well as education as a means of preventing AIDS.

13. Urban environments (option G)

This optional theme considers cities as places of intense human interaction. Urbanization promotes efficiency through economies of scale and concentration of the workforce. Towns and cities are centres of economic activity, administration and the hubs in a communications network, sometimes globally.

Cities are socially and culturally diverse and many possess distinct ethnic districts and stark contrasts in wealth, which may result in inequities and crime, especially in large megacities where populations may be on the move and surveillance systems weak. The rapid rate of urbanization and high levels of consumption present challenges to urban planners. When the pace of urbanization exceeds the rate at which it can be successfully managed, cities suffer from various stresses. Environmental stressors include air pollution, the urban heat island effect and the excessive production of waste.

Overcrowding and poor living standards are typical of many cities in the poor world although sociological structures may be sound and worth maintaining, along with the informal economy that sustains them. The destruction of inner-city slums is controversial because it will disrupt families and business networks. These poor districts also offer services such as laundry and recycling, which provide important income. Cities in the rich world are not exempt from problems, although these tend to be on a much smaller scale. The focus of urban planning in the rich world has been to revive the inner city and the central business district, which has suffered from the centrifugal movement of populations and businesses.

The future survival of cities depends upon the sustainable management of resources, the minimization of waste and the eradication of social inequality.

Urban populations

You should be able to:

- define urbanization and explain the variation in global growth rates and patterns.

Urbanization is the process whereby an increasing percentage of a country's population come to live in towns and cities. No other definition is acceptable; the physical expansion of the city's built-up area is urban sprawl, not urbanization.

The level of urbanization is highest in more economically developed countries, but the growth rate has slowed down or ceased owing to very low rates of natural increase and increasing counter-urbanization. In the less-developed world, the level of urbanization is lower, but the rate is increasing rapidly because of rural–urban migration.

You should be able to:

- explain the processes of centripetal movement (rural–urban migration, gentrification, re-urbanization/urban renewal).

Explain centripetal movement by mentioning the push and pull factors that attract and repel people to and from cities and towns. The situation is dynamic and has changed markedly in the last 50 years, with a shift from centripetal to centrifugal movement (the processes of suburbanization and counter-urbanization). Latterly, urban renewal has revived some city centres and re-urbanization has occurred.

You should be able to:

- explain the process of centrifugal movements (suburbanization, counter-urbanization, urban sprawl).

Explain centrifugal movement by mentioning the push and pull factors that attract and repel people to and from cities and towns. The situation is dynamic and has changed markedly in the last 50 years, with a shift from centripetal movement (the processes of rural–urban migration, gentrification, re-urbanization/urban renewal) to centrifugal movement. Latterly, urban renewal has revived some city centres and re-urbanization has occurred.

You should be able to:

- explain the contribution of natural change to patterns of population density within urban areas.

In more-developed cities, high population densities are common in inner-city districts that are often occupied by poor, young migrants from ethnic minorities, and with high birth rates. These districts have high-density housing and are often overcrowded. In contrast, the suburbs are usually occupied by affluent older people, giving these districts a low rate of natural increase.

You should be able to:

- explain the global increase in the number and location of megacities (population over 10 million).

Rapidly growing cities are found in the less-developed world and are characterized by net immigration and high rates of natural increase related to their youthful populations. Megacities have developed because of an underdeveloped urban hierarchy with a focus on one major, often capital, city, which can have enormous attraction over a very wide hinterland. Megacities have a prestige that draws in foreign capital and, often, foreign populations. The prospects for work there are more hopeful than in smaller urban areas.

Urban land use

You should be able to:

- explain the location of residential areas in relation to wealth, ethnicity, family status (stage in life cycle)
- examine patterns of urban poverty and deprivation (such as slums, squatter settlements, areas of low-cost housing and inner-city areas
- examine the causes and effects of the movement of socio-economic groups since the 1980s.

Be prepared to sketch the town or city showing neighbourhood contrasts in wealth, health and ethnicity. Knowledge of your local urban area is particularly useful, especially if you know it in detail and can refer to named places within the town or city.

Rapidly growing cities tend to have poor immigrant populations living in squatter settlements on the edge or in areas of disamenity towards the centre. In more-developed cities, poorer and younger populations tend to be close to the centre and the older and wealthier populations further towards the suburbs.

There is a tendency for migrants into the city to occupy the cheapest residential areas close to the centre or, in the case of the less-developed world, in squatter settlements on the periphery. In time, many urban populations seek to move away from these areas of high housing densities in poor conditions, provided their income supports it. They move towards the suburbs and can sometimes maintain their ethnic concentrations there. The movement of ethnic minorities out to the affluent suburbs can cause friction with local residents. Similarly, the re-urbanization process has attracted younger, upwardly mobile couples who now occupy many of the rejuvenated and desirable residences in inner-city areas, for example, in Detroit. The displacement of local residents has been a source of great friction.

You should be able to:

- explain the spatial pattern of economic activity, the zoning of urban and suburban functions and the internal structure of the central business district (CBD)
- describe the informal sector; its characteristics and location in urban areas
- examine the causes and effects of the movement of retailing, service and manufacturing activities to new locations, including brownfield sites.

Knowledge of your nearest large city, or one that you have surveyed, would be particularly useful here. Be prepared to draw a plan showing the location of industrial areas, the formal CBD and the location of informal activity. Ideally, you should be able to sketch the commercial land uses within the CBD itself.

Informal activity can take many forms and includes retailing, service and manufacturing industry. It is characteristic of the less-developed world, where it dominates the economy, but is also found elsewhere in more-developed countries. Whatever the location, informal activity involves face-to-face transactions, little paperwork and the minimum administrative recording. Be prepared to describe its advantages and disadvantages relative to the formal economy.

Urban stress

You should be able to:

- examine the effects of structures and human activities on urban microclimates, including the urban heat island effect and air pollution.

Show an understanding of the factors that are responsible for the creation of an urban microclimate, such as urban structures, materials and anthropogenic heat sources. Remember that the intensity of the urban heat island varies in time and space, and that seasonal and diurnal patterns should be considered.

You should be able to:

- examine the other symptoms of urban stress including congestion, overcrowding and noise, depletion of green space, waste overburden, poor quality housing, social deprivation, crime inequality.

To examine these other symptoms, you will need to consider their causes and impacts with reference to at least one example of a named city. Many large world cities will provide this information.

The sustainable city

You should be able to:

- describe the city as a system in terms of:
    - inputs—energy, water, people, materials, products, food (urban agriculture)
    - outputs—solid, atmospheric and liquid waste, noise, people
- distinguish between a sustainable circular system where inputs are reduced and outputs are recycled and an unsustainable (open/linear) city system with uncontrolled inputs and outputs.

In addition to a written description, you should be prepared to draw labelled or annotated diagrams illustrating the two systems of resource flow through the cities. Refer to chapter 5, "Data response skills", for advice on sketching and labelling a diagram.

Be aware of the command term "distinguish". This means commenting on the differences factor by factor, in a systematic way. You should not describe one system and then the other.

You should be able to:

- referring to at least **two** city case studies, discuss the concepts of:
  - sustainable city management
  - the urban ecological footprint.

Sustainable city management aims to meet the needs of the present generation without compromising the needs of future generations. This means that the city's population consumes relatively low quantities of energy, water and other resources, its transport systems are efficient and its population has social and economic security.

The urban ecological footprint is the land area required to supply the city with natural resources and absorb its wastes. You should be able to discuss this concept and support your answer with a range of footprint data to represent cities of contrasting sizes and levels of development. Curitiba in Brazil is a good example of a sustainable city.

You should be able to:

- evaluate **one** case study from **each** of the following.
  - One socially sustainable housing management strategy.
  - One environmentally sustainable pollution management strategy.
  - One strategy to control rapid city growth resulting from in-migration.

You will have already chosen two case studies to exemplify the concept of the sustainable city and it's possible that one or both may be relevant to the three urban management strategies listed above. Some good examples are: housing management in Mumbai and Singapore; waste disposal in Kathmandu; rapid city growth in Shanghai. In each case, you must be prepared to write an evaluative answer that discusses and weighs up the effectiveness of each strategy.

(a) Describe what is meant by a "sustainable city". *[4 marks]*

(b) Explain **three** ways in which human activities can modify the microclimate of an urban area. *[3×2 marks]*

(c) Examine reasons why cities in some parts of the world have higher rates of population growth than others. *[10 marks]*

[Taken from paper 2, May 2011]

How do I approach the question?

(a) A diagram is not asked for, but would certainly help this answer.

(b) Be aware of the term "microclimate". This means temperature, humidity, wind speed, precipitation and clouding. The greenhouse effect is irrelevant.

(c) City population growth needs explaining at the start of your answer in terms of the relationship between natural increase (excess births over deaths) and a positive migratory balance (more people moving in than out). This question does not require a discussion of urban sprawl (the physical expansion of the city population). Remember to look at the global variation in urban population growth rates and explain them. Including statistics for annual percentage growth, life expectancy and fertility rates is important. Examples of cities that show contracting rates are essential.

What are the key areas of the syllabus?

- Option G: Urban environments
- Sub-topics: Urban populations; Urban stress

This answer achieved 16/20

(a) A sustainable city is a circular metabolised city:

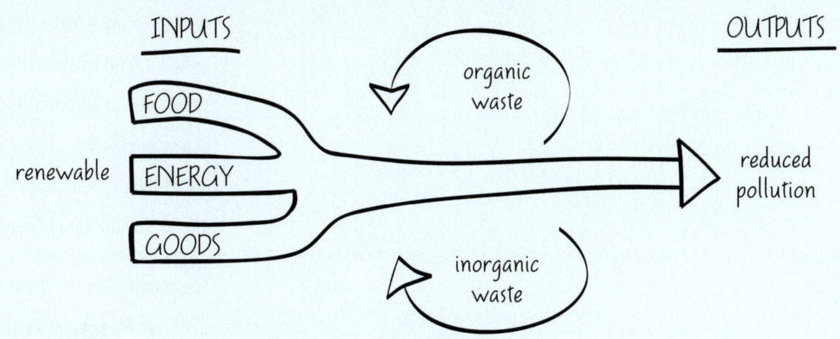

The diagram is helpful in defining a sustainable city. The control of resource consumption and waste production are mentioned, but slightly more detail is needed.

They will use more renewable energy resources and recycle, re-use and manage wastes much better organic and inorganic wasted. they will also then reduce emissions and pollution.

They will therefore not harm the environment (in the same way). Different strategies can be implemented, e.g. Hoy no Circula in Mexico, reducing emissions by 17% the first year. [3]

(b) Humans can modify the microclimate of an urban area by:

(1) Setting up buildings (tall or short) Tall buildings will give cooler temperatures at street level, as not as much radiation from the sun is able to reach it. Short buildings will heat up the street level:

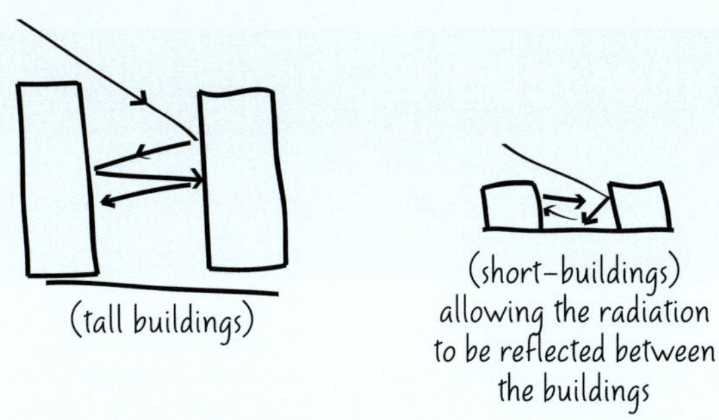

As well as this, buildings may reduce wind speeds (so its warmer or channel them (so it's very windy).

(2) They can also build/use much materials with a lower albedo (lower reflectivity) which will absorb more heat which will then be released at night (urban – heat – island effect), (included in that, rougher surfaces as well, causing also more reflection, and warmer temperature.

(3) They will add anthropogenic heat to the atmosphere through concentrated road traffic emissions, central heating and air conditioning. [6]

Three activities are identified and clearly explained.

(c) Cities in some parts of the world have higher rates of population growth than others.

One can compare MEDC's and LEDC's for instance, and the size of the cities, the structure of the cities (poorer regions eg. shanty towns).

Global differences in urban population growth rates are recognized and fully explained. However, statistics for relative growth rates are needed.

MEDC cities seem to grow but not as fast or as big LEDC cities. This may be due to a lot of rural – to – urban migration still occurring how rapidly in developing countries, whereas in MEDC's we have already reached the maximum, and we are even seeing counter – urbanisation occurring. As well as this, in many LEDC's they have higher growth rates, and natural increase due to higher fertility rates, perhaps due to the lower status of women in these poorer countries or other socio – economic factors. Urban sprawl and suburbanisation is also increasing at a bigger rate in LEDC's as distances are becoming greater, car ownership is increasing.

Looking at the size of cities, we can also compare with megacities (a population of more than 10 million people). Here, we still see many industries located and moving there, as it has a good large market, good access to international markets and much labour available this means with the industries, people will also make for work there. Moreover, megacities, with already a big population will have a greater natural increase than smaller cities, due to more people, therefore a higher birth rate. They also attract many people due to all things available there. Looking at the structure of cities, we see that cities with a greater young population usually have higher rates of population growth. More young will move in, and they are in the pre-childbearing ages or child bearing ages. These cities can have a large youthful population due to perhaps large/many shanty towns. An example is the favelas in Rio de Janiero.

The people living in them are estimated around 2.5 million (out of the 12 million living there) ie. About 20% of the population there! Here we see higher population growths. This can even be compared between MEDC's and LEDC's, as we often find more and larger shanty towns in LEDC's. Of the urban population, in MEDC's shanty towns make up 32% whereas in LEDC's they make up 78% of the total urban population!

Some cities we also see how pollution has increased so much that it becomes less attractive to people, so less people move in there, and we may perhaps see more centrifugal movements (eg. counter – urbanization, urban sprawl, suburbanization etc) This can also be due to for instance very large crime rates in certain big cities compared to others. [7]

Examiner report

The answer focuses well on the question and shows a good appreciation of urban population dynamics. More might have been said about the near negative rates of natural increase in MEDC cities.

This answer achieved 8/20

The student focuses upon only two aspects of sustainable urban management, but more might have been said about the specific aspects such as city design, inputs of water and energy, and effectiveness of the infrastructural system.

(a) A sustainable city is a city that reduces waste from input through recycling and reuse. Recycling is desired as the reuse of industrial and household waste by being sent to manufacturer, these two methods reduce waste that can affect the environment and create economic problems in trying to fix it, this trend is seen in Stockholm where constant recycling mechanisms insure sustainability in resource consumption, input and evidently output. [2]

The student makes some valid references to wind modifications but shows some confusion over the effects of albedo and greenhouse gases.

(b) The air around activities can be drier and more dust concentrated as a result of lack of vegetation and numerous draining mechanisms. The lack of vegetation also could lead to less albedo effect where absorption of heat is noticed by the increased use of heat absorbing materials in buildings and other infrastructures. The increased amount of perpendicular structures could in turn affect wind movement and could cause temperature differences on buildings and on level streets, tunnel wind mechanisms occur when street is aligned to direction of local winds, making areas more windy. higher expulsion of green house gases, pollutants and hygrosepic particles make area warmer as a result of the trapping of long-wave radiation. Therefore rising temperature, drier air, windy trends and variation in temperature within different location of cities are created as a result of human activities. [3]

(c) There are different reasons as to why population increase is noticed in exponential rates in LEDC's than in MEDC's. This is a result of different social, political and economical factors that are playing apart in different societies.

One factor is the status of women. It has been noticed that extensive social pressure on women in rural areas, the disparity in educational and economical opportunities has a positive effect on the growth of population. The increasing economical and social isolation of women leads to more bigger family sizes. In rural areas of Ethiopia where status of women is low and education opportunities are also low, families tend to have 4 more kids than Ethiopians with an educational background. The economic benefits of having more children in rural areas also plays a part in this where more kids can sustain their parents at old age and can help on the farm. However the role of education is also in urban areas like Estonia where uneducated families have 3 kids while educated families have an average of 1.5. The status of women is crucial in this because if she doesn't have a say in the size of the population it can lead to her having more children. This is one of the many reasons why empowerment of women in the millennium development goal as goal number 3. Elevated status of women in MEDC's such as Sweden shows that women in Sweden have an average of 1.2 kids.

The availability of healthcare and its easy access also provides better maternal health which reduces infant mortality and the need to compensate for lost children. Therefore the status of women, the access to education and economic opportunities have a direct effect on the growth of a population of an area.

Other causes of exponential increase in population growth could be the standard of living in an area and the actual demographics of the population. Economic prosperity promotes population growth because having children can be seen as affordable, however the U.N argues that by raising the standard of living in LEDC's women can also participate in the job industry to increase economic flow and this inturn leads to the emancipation of women and their rise of their status. This is why the U.N argues countries like Kenya to increase their standard of living through equal distribution of resource among the population and decrease in corruption.

The demographics of a population also is an important factor. If there are more people that are close or are at child bearing age than those who are dependant of them, population growth can occur in a short amount of time. This is seen in LEDCs like Ethiopia where the life expectancy is around 68 meaning that the population is less affected by death rate. This is not the case in MEDCs like Sweden where life expectancy is +75 and 7% of the population demographic is of this portion of the society, creating less annual population growth.

Therefore population growth is seen more in LEDC than MEDC's as a result of low status of women. Lack of education, low standards of living and economic benefits and much higher younger population than in MEDCs. [3]

Examiner report

The focus is almost entirely upon the status of women and the differences in fertility rates between rural and urban areas. There is no reference to the relationship between fertility and mortality, which determine differential rates of metropolitan natural increase between the more-developed and the less-developed worlds. Knowledge of specific growth rates is very thin and there are no references to MEDC cities.

14. Global interactions (HL students only)

Global interaction suggests a two-way and complex process where cultural traits and commodities may be adopted, adapted or resisted by societies. The process is neither inevitable nor universal.

Measuring global interactions provides an introduction to the course by identifying the pattern, level and rate of global interactions. This is not a "globalization" unit—but globalization is an important global interaction.

You should be able to cover and illustrate all topics by the use of case studies and local examples where relevant.

Measuring global interactions

You should be able to:

- describe and evaluate one of the following two globalization indices: the AT Kearney index or the KOF index, as a measure of global interaction
- describe how the globalization index may be represented spatially.

"Evaluate" is a higher-order command term. You are expected to make a judgment by weighing up the strengths and limitations of one of the indices. For example, are all the components that make up the index equally weighted?

Indices can be shown in map format (as in the case of the KOF index) or as a series of compound bar charts (as in the Kearney index). There is no requirement to do both.

You should be able to:

- discuss the spatial pattern of global interactions through the mapping of core areas at the focus of interaction (network hubs/nodes), the peripheries and areas relatively unaffected by these interactions.

There are a number of ways of describing cores and peripheries. The most frequently used is Wallerstein's world-systems analysis, which divides the world into a core, semi-periphery and periphery. Friedmann's world city hypothesis divided the major cities of the world into core and semi-periphery, each containing primary and secondary cities.

How do I approach the question?

- **Core and periphery**: The concept of a developed core surrounded by an undeveloped periphery. The concept can be applied at various scales.
- **Globalization**: "The growing interdependence of countries worldwide through the increasing volume and variety of cross-border transactions in goods and services and of international capital flows, and through the more rapid and widespread diffusion of technology" (source: IMF).
- **Globalization indices**: The AT Kearney *Foreign Policy* index measures twelve variables, which are subdivided into four "baskets": economic integration, personal contact, technological connectivity and political engagement. Nations are ranked according to a calculated globalization index.

  The KOF index measures three main dimensions of globalization: economic, political and social, and nations are ranked accordingly. It is designed by the Swiss Federal Institute of Technology on a yearly basis.
- **Node**: The main centre or focal point.

Changing space—the shrinking world

Changing space—the shrinking world identifies improved information and communications technology (ICT) and transport as fundamental to all forms of global interaction. It examines the pattern(s) and process(es) of global interactions and the technology that has enabled them.

You should be able to:

- explain how a reduction in the friction of distance results in time–space convergence
- examine the relative changes in the speed and capacity of two types of transport (air, ocean, road, rail, pipeline) responsible for the flow of goods, materials and people.

You should give a detailed account of how forms of transport have changed over time, and allow people and cargo to travel longer distances in less time. This brings places closer together.

To achieve the best results, you should choose two contrasting forms of transport that have very different advantages in terms of transporting goods, materials and/or people, and show how the relative speed and capacity of each has changed over time.

You should be able to:

- examine the changes in a transport, internet or telecommunications network in terms of the extension of links and nodes and the intensity of use at a national or global scale
- describe the role of information and communications technology (ICT) in civil society and the transmission and flow of images, ideas, information and finance
- examine the contrasting rates, levels and patterns of adoption of an element of ICT in two countries.

Examples of changes include the evolution of the Indian rail network since the nineteenth century and the recent growth of the mobile phone network in Africa.

Improvements in ICT have facilitated the spread of information, ideas and finance around the world.

Although there is increasing access to ICT among people, there are still very different rates of access to ICT between countries, for example, the number of computers per 100 people in the USA and India.

Key definitions

- **Civil society**: Any organization or movement that works in the area between the household, the private sector and the state to negotiate matters of public concern. Civil society includes non-governmental organizations (NGOs), community groups, trade unions, academic institutions and faith-based organizations.
- **Diffusion**: The spread of a phenomenon over space and through time. Diffusion can either be to areas close by or can be hierarchical, for example, from larger cities to smaller cities and then to towns.
- **Frictional effect of distance (distance decay)**: The way in which places that are closer to each other interact more closely whereas places that are further apart interact less. Time–space convergence leads to a reduction in the frictional effect of distance.
- **Time–space convergence**: The reduction in the time taken to travel between two places due to improvements in transportation or communication technology.

Economic interactions and flows

You should be able to:

- examine the importance of loans, debt repayment, development aid, remittances, foreign direct investment and repatriation of profits in the transfer of capital between the developed core areas and the peripheries
- examine the influence of governments, world trading organizations and financial institutions (such as the World Trade Organization, International Monetary Fund and World Bank) in the transfer of capital.

There are a number of ways in which capital is transferred from the core to the periphery. India is the largest recipient of remittances in the world, and remittances may exceed the total value of international aid that is given to countries. Nevertheless, there is also some transfer between parts of the core and from the periphery to the core.

The following bodies influence the transfer of capital.

The World Trade Organization (WTO) deals with the rules of trade between nations at a global or near-global level. The WTO includes trade in services, creations and designs (intellectual property). WTO agreements provide the legal ground rules for international commerce, which aim to help trade flow as freely as possible.

The International Monetary Fund (IMF) is the organization that oversees the global financial system by following the economic policies of its member countries, in particular those with an impact on the exchange rate and the balance of payments.

The World Bank's aims have evolved over time. Initially it lent relatively small amounts of money to countries, then increased its lending in the 1970s and 1980s. It then became involved with the creation of structural adjustment policies and, most recently, has been engaged with the United Nations' Millennium Development Goals.

You should be able to:

- explain the causes and effects of one major flow of labour between two countries.

Remember that international migration was also considered in the core (see "Populations in transition"). If the international migration you studied was a labour flow, you can use it here. Most labour flows are either within the migrant's home region or to rich countries.

You should be able to:

- explain the role of ICT in the growth of international outsourcing.

ICT has facilitated the level of international outsourcing by connecting more parts of the world. Different locations offer varied opportunities (skilled labour, business environment, financial structure) for outsourcing.

Key definitions

- **Millennium Development Goals**: The United Nations' global action plan to adopt eight anti-poverty goals by 2015.
- **Outsourcing**: The concept of taking internal company functions and paying an outside firm to handle them. Outsourcing is done to save money, improve quality or free company resources for other activities.
- **Remittances**: Transfers of money/goods by foreign workers to their home countries.
- **Structural adjustment policies**: Conditions imposed by the World Bank and the IMF on countries to borrow aid in return for reducing government spending in certain areas such as health and education.

Environmental change

You should be able to:

- identify the effects of agro-industrialization and changes in international production and consumption on the physical environment
- discuss the environmental consequences of increasing international demand for one raw material
- examine the concept of food miles and the environmental consequences of increasing volumes of air freight.

The impact of agro-industrialization is producing large quantities of food at ever-competitive prices. However, it is also leading to a reduction in biodiversity, using up scarce water resources and putting the demands of rich countries ahead of the needs of poor countries. Increased demand for biofuels is resulting in destruction of tropical rainforests, reduced biodiversity, increased run-off, siltation and flooding.

Increased international demand for oil is leading to oil exploration in ever-more hostile environments. (There is overlap here with the core theme and with option B—oceans and their coastal margins.)

A typical UK Christmas dinner may transport food over 37,000 km. Transport accounts for about 25% of the world's CO_2 discharge and, of that, 13% is produced by aviation.

You should be able to:

- discuss the reasons for and consequences of the relocation of polluting industries (such as some TNCs) and waste disposal (such as ICT, chemical and nuclear waste) to countries with weaker environmental controls and safety regulations.

Many newly industrialized countries (NICs) have relaxed environmental legislation. The *macquiadoras* industries in Mexico are a good example—industrialists have been attracted by low labour costs, good access to the US market and weak environmental legislation.

You should be able to:

- describe **one** major pollution event affecting more than one country and examine the consequences of and responses to this event
- examine the growth of environmental awareness as a consequence of these global interactions
- examine the role of **one** international civil society organization in fostering improved environmental management.

Acid rain remains one of the most notable forms of transboundary pollution. It is increasingly a problem in parts of Asia, as a result of industrialization in China and India. Dust from the Sahara has been linked with increased asthma attacks in the Caribbean. Greenpeace and the World Wide Fund for Nature (WWFN) are good examples of international civil society.

You should be able to:

- explain the evolution of uniform urban landscapes; the effects of common commercial activity, structures, styles of construction and infrastructure.

The spread of McDonalds, Starbucks and IKEA are all good examples of how cities are becoming homogenized. The globalization of economic activity and culture (for example, art, media and sport) are all ways in which cities are becoming similar in appearance. Increased migration, tourism and the time–space convergence allow ideas and products to be more widely available in cities of different parts of the world.

Key definitions

- **Agro-industrialization**: Commercial farming that is carried out on a large scale, with large inputs of mechanization and capital. It aims to maximize output.
- **Food miles**: A measure of the distance food travels from its source to the consumer. This can be given either in units of actual distance or of energy consumed during transport.
- **Homogenization of urban landscape**: The process by which urban areas are becoming more similar in terms of the companies that locate there, their appearance and the cultural experiences they offer (Chinese restaurants, Italian pizzerias, Indian restaurants, films in cinemas, etc).
- **Transboundary pollution**: A type of pollution where the cause of the pollution is in one country but the effects are felt in another, for example, acid rain. There have been very small amounts of fall-out in the USA following the Fukoshima nuclear power disaster in Japan in 2011.
- **Transnational corporation (TNC)**: A firm that owns or controls productive operations in more than one country through foreign direct investment.

Sociocultural exchanges

You should be able to:

- describe cultural traits in terms of language, customs, beliefs, dress, images, music, food and technology
- examine the diffusion of cultural traits resulting from the international movement of workers, tourists and commodities.

The spread of the English language and the game of cricket are good examples of cultural traits. Equally, the popularity of Indian and Italian foods shows a diffusion of cultural traits.

The growth of British-style pubs or Irish-themed pubs and supermarkets in tourist resorts, such as in Playa de las Americas in Tenerife, is another example of cultural diffusion. Growing Chinese influences in parts of Sub-Saharan Africa is an excellent contemporary example.

You should be able to:

- describe the role of TNCs and the media in spreading consumer culture
- select **two** different branded commodities and examine the spatial and temporal pattern of adoption on a global scale.

McDonalds, Coca-Cola, Starbucks and IKEA are all very good examples of branded goods that have spread globally. The speed of diffusion has accelerated in recent decades.

You should be able to:

- examine the role of diasporas in preserving culture in **one** country and the adoption of minority traits by host societies
- examine the impact of cultural diffusion on **one** indigenous and remote society through the influence of international interactions
- examine the ways in which international interactions may result in the homogenization and dilution of culture
- define and exemplify the concept of cultural imperialism.

The Irish in the USA are a very important and powerful diaspora. Irish culture has been maintained in terms of music, dance, language, Gaelic football, hurling and Irish food and drink. Many of these have been popularized and adopted by the host population, for example, St Patrick's Day parades in New York, Chicago and Boston.

Many indigenous groups have been adversely affected by contact with Western civilization. The Jarawa of the Andaman Islands have, for example, been affected by poachers, Indian settlers, new tourism developments and disease.

International movement of workers, tourists and commodities makes places more similar. The spread of global brands, such as Disney and McDonalds, has lead to a global culture and the dilution of local traditions.

Cultural imperialism occurs when a larger, more powerful nation promotes its culture in another, less powerful country, for example, the spread of the Chinese culture in Tibet.

Key definitions

- **Consumer culture**: A society that places great emphasis on the consumption of bought items.
- **Cultural imperialism**: The practice of promoting the culture/language of one nation in another. It is usually the case that the former is a large, economically or militarily powerful nation and the latter is a smaller, less affluent one.
- **Diaspora**: A migrant population living abroad, for example, the Irish in the USA, the current Irish migration to Australia or the Polish migration to the UK.
- **Homogenization**: The process of becoming similar.

Political outcomes

You should be able to:

- discuss the links between the diminishing effectiveness of political borders and the flow of goods, capital, labour and ideas, and the role of **one** multi-governmental organization such as the European Union (EU), the Association of Southeast Asian Nations (ASEAN), the North American Free Trade Agreement (NAFTA) and MERCOSUR (the common market of South America)
- discuss the shift of power from nation state to TNCs as a result of their economic size and dominance
- compare the wealth of TNCs with that of nation states.

Member countries of the EU have given up some of their sovereignty in return for greater market access and a say in the political power of the expanded EU.

Some TNCs, such as Wal-Mart, Ford, ExxonMobil and Royal Dutch Shell, have assets far in excess of some poor countries. However, TNCs are not all-powerful, as seen in the case of BP over the 2010 Deepwater Horizon oil spill disaster.

You should be able to:

- examine the resurgence of nationalism in one country as it attempts to retain control of its resources and culture
- discuss anti-globalization movements
- discuss the attempts to control migration into one country.

The resurgence of nationalism can be seen in a wide range of countries including Nunavut (Canada), states of the Former Yugoslav Republic (FYR) and the Former Soviet Union (FSU).

Anti-globalization movements vary from a national level, such as the rejection of the euro currency by the British and Danish governments, to action groups, such as the People's Action Group.

Migration control is illustrated by international border controls, for example, in the USA and Australia.

Key definitions

- **Anti-globalization**: Individuals, organizations and governments that may resist the spread of globalization, for example, North Korea's government and Peoples' Global Action.
- **Multi-government organization**: A group of countries, such as the European Union, in which the membership is made up of two or more countries.
- **Nationalism**: Principles of patriotic feelings leading to a desire for self-government and control of its own resources and culture, but which can include acts of terrorism.
- **Sovereignty**: Having the right to rule and govern a defined territory.

Global interactions at the local level

You should be able to:

- distinguish between the terms **globalization** and **glocalization**.

Globalization is "the growing interdependence of countries worldwide through the increasing volume and variety of cross-border transactions in goods and services and of international capital flows, and through the more rapid and widespread diffusion of technology" (source: IMF).

Glocalization is the adaptation of a good for a local market. The adaptation of McDonalds to the Indian market (with the chicken maharaja mac) is a good example, as is the example of small Japanese cars for the European market and sports utility vehicles (SUVs) for the North American market.

You should be able to:

- examine the extent to which commercial activities at a local scale have become globalized
- examine the reasons why the level and rate of adoption varies from place to place.

In many parts of the world, commercial activities have become globalized. Barclays, Tesco, Marks and Spencer, McDonalds and Starbucks, for example, have all become globalized.

The rate of adoption varies with affordability and acceptability—some countries may resist globalization, for example, Myanmar and North Korea.

You should be able to:

- discuss civil society responses to globalization; the adoption, adaptation (glocalization) or rejection of globalized goods, services and cultural traits
- evaluate the relative costs and benefits of local commercial production to the producer, the consumer and the local economy, compared with the costs and benefits of globalized production.

Local responses can lead to a product being adapted, for example, Wal-Mart varied its style to fit the Chinese market and the German market.

Costs and benefits can be seen with regard to the food industry. Mass-produced foods may be cheap, but they can be environmentally damaging. Local farmers' produce may be environmentally friendly—although this is not always the case—but generally costs more.

You should be able to:

- describe the role of civil society in raising awareness of local and global environmental, social and cultural issues
- examine the role of civil society in supporting local economic activity and strengthening local cultural values
- discuss the position held by anti-globalization groups
- evaluate the quality of life of a contemporary non-globalized society.

A good example is the role of Greenpeace, the Body Shop and Friends of the Earth in raising awareness over Shell's involvement in the Niger delta with the Ogoni population.

Avaaz is an NGO that is involved in political campaigns largely via the internet. Survival International is an NGO that helps raise awareness about indigenous groups.

For some non-globalized societies, such as the Popaluca Indians of Mexico, their traditional ways provide a high quality of life, but this is being threatened as their forests are destroyed.

Key definitions

- **Cross-border transactions**: Sales of goods or services between different countries.
- **Glocalization**: A term that was invented to emphasize that the globalization of a product is more likely to succeed when the product or service is adapted specifically to each locality or culture in which it is marketed. The increasing presence of McDonald's restaurants worldwide is an example of globalization, while changes made to the menus of the restaurant chain, in an attempt to appeal to local tastes, are an example of glocalization.

(a) Explain how and why **one** network (transport, internet, or communication) has grown over time. *[10 marks]*

(b) Examine the relative importance of the different financial flows that connect global core areas with peripheral areas. *[15 marks]*

[Taken from paper 3, higher level, May 2011]

How do I approach the question?

(a) The question asks you to explain both how and why one network has grown over time. Therefore, there are two distinct aspects—how and why. You will only be credited for referring to one network—transport, internet or communication. The "how" part of the question should include changes in user numbers, and changes in the global pattern of distribution. The "why" part should consider the rising number of users, new markets, the role of TNCs, and changes in technology. To access markband D, both "how" and "why" must be addressed. To achieve markband E, your answer should be well balanced.

(b) This question wants you to examine the relative importance of different forms of financial flows that connect global core areas with peripheral areas. Therefore, you must identify different forms of financial flows, such as loans, foreign direct investment (FDI), profit flows, remittances and international aid. You should also define the global core and periphery. Note that the scale required is global, not national or regional.

The question wants you to reach an answer—what is the most important financial flow? It is not enough just to describe them—their relative importance needs to be assessed.

What are the key areas of the syllabus?

- HL extension: Global interactions
- Sub-topics: Changing space—the shrinking world; Economic interactions and flows

Note: These are synoptic essays so information from all parts of the syllabus can be relevant.

This answer achieved 11/25

The student has made a general statement about transporting people and goods. It would be useful to say by how much population has grown in the twentieth century, and identify the major trade routes.

In this paragraph, the student identifies a range of reasons for movement of people— such as unemployment and tourism— and movement of goods.

The student addresses "why" and "how".

If the student had referred to specific travel times—time-space convergence—or referred to specific innovations—jet aeroplanes or airbus—the answer would have more detail.

This answer focuses mainly on "why"—treatment of the "'how" is superficial. The response is not very balanced and lacks detail. No real supporting case studies are referred to, therefore the student only partially answers the question—markband C.

The student defines the core and periphery, and links the two.

The student identifies some benefits for the periphery as a result of interactions with the core but does not specify which financial flows.

The student does not engage directly with the question—different financial flows have not been addressed yet.

(a) Since the 20th century, the transport network has grown over time by means of quality and quantity. With the rapid growth of global population, the need for more transport networks increased. Despite the growth of global population, globalization is another factor which has made possible long-distance transport. This network is not only used to transport people from one place to another, but it is even used to transport goods.

With the increasing need for faster deliveries, people have tried to make the transport network more efficient, better quality, safer and quicker. Due to globalization, people find it easier to travel from one country to another as the formal barriers between countries have been reduced. High rates of unemployment in a particular country had resulted in people wanting to emigrate and using more transport networks. With the development of tourism, the transport network has grown over time, with the more interaction level between countries in terms of trade, politics, economic, etc has increased the need of people moving from one place to another.

By inventing new ways of transport, people have made it easier to trade long-distance in shorter time. With the invention of trains, airplanes which are produced faster and faster speed, has made the transport network grow over time.

Due to globalization, countries interact more with each-other and communiticating with each other is a main need. Due to the technology, resources are used more as it is easier to produce food for example in large amounts so, this amount of food has to be distributed to the rapid growing population not only locally and nationally, but being transported overseas. So, the importance of transport network has grown a lot over time and especially recently. [5]

(b) The core areas are called those developed countries and the developing countries are called the peripheral areas. However the core areas and peripheral areas interact politically, economically, socially, etc. In other words, the core peripheral areas need each other to survive.

Peripheral areas, having higher rate of unemployment, less developed technology, low income, more poverty, would benefit a lot if they would interact with the core areas as they are able to invest in the peripheral areas. The core areas would give the peripheral areas the technology, political empowerment etc which would be very beneficial for the developing countries. The importance of the different financial flows that connect global core areas with peripheral areas is that the developed countries use the very cheap labour that exists in the peripheral areas and cheap resources too which the local communities may not have the technology to make use of the energy, food resources available. So, the developed countries bring money to the developing countries by investing or importing the goods from the peripheral areas to their own country. Instead, the peripheral areas benefit and the core areas may make use of the cheap labour and reduce unemployment rates in peripheral areas.

The student still refers to financial flows in a very general way—different financial flows have not been identified or discussed.

Emigration is mentioned but nothing is said about remittances or their value.

However, the different financial flows may not be spread to the whole population of peripheral areas and many people may still live in poverty or having deadly diseases. In the same time, the different financial flows have helped in reducing disparities between places located in peripheral and core areas.

As formality barriers between countries have reduced recently and people find it easier to emigrate, people tend to emigrate from the peripheral areas to core areas because of better chances of employment with higher wages too and the opposite when people from core areas make use of cheap labour in peripheral areas by minimizing their costs and maximising profit.

Many powerful international companies tend to produce their products in a less developed country and sell it at a higher price to more developed country. Still, the peripheral areas benefit from the availability of new jobs, new schools built, hospitals, technology available. [6]

Examiner report

The student's work is not very detailed. There is a limited range of financial flows and much of the material is implicit. There is limited support/use of case studies. Many of the statements are very generalized. The answer only partially answers the question—markband C.

This answer achieved 25/25

Good focus on transport and use of terminology—a brisk opening.

The student looks at "why" and compares case studies. The student makes the case about why transport is important and how it aids development.

A topical answer. The student uses case studies that are very recent.

Good on "why"—the student uses data for Kenyan horticulture and agriculture to support the point.

(a) Transport networks have grown over time, and this was illustrated by Janelle, who mapped London to Edinburgh journey times since the 1800s, and described the "space–time convergence". Over time, the world has become more connected, and transport is key to allowing this connectivity.

LEDCs governments know that by having better transport links they will become more accessible, and so will attract more foreign direct investment (FDI). This FDI is key to their growth and development. India is attractive for FDI due to its huge market, low cost but educated workforce, but also due to the well connected railway system: over 60,000 km of railway tracks across India. Compare is to the limited railway network in Africa, where FDI is less attracted. It can be seen that transport networks there were driven by imperial ambition and exploitation of resources: the east is well connected but the west and central areas are not, and the Tan-Zam railway to transport copper from mines is an example of imperial exploitation creating a need for transport.

The increasing freedom of movement over time has caused cheaper methods of transport to flourish: since the EU allowed freedom of movement, Eurostar, Ryanair and Easyjet have flourished. Similarly in India the increasing affluence of the middle classes has created a need for low cost travel: Go Airlines. This is the time-space convergence: areas have become more accessible, and by more people. The sky train into Tibet, and allows the Han Chinese to move into Tibet with ease.

The importance of good transport links is clear: Japan has given aid to India for the construction of a mass rapid transport system on Delhi: subways, elevated roads, and a better road network. However, transport has also improved due to a need for resources and the increased amount of offshoring. Post 1945, container shipping, mass movement of freight on cargoships and lorries have minimised break of bulk points through more efficient transport fuelled by capitalism and free trade. This extended to flying fresh produce and 'food miles' later in the 20th century: Kenya flies 1 million kg of fresh produce each night as exports of flowers and vegetables, ¾ of which is exported to the EU.

Therefore, transport has grown over time due to the requirement of infrastructure to allow globalisation; movement of people, resources and goods. [10]

A focused start. The student links core and periphery although financial flows are not developed as yet.

(b) The idea of a global core area is of an area that is particularly well connected: 'world cities' such as Stanford, Sao Paulo or London. These then diffuse into the periphery areas. Innovation and ideas, good, technology and investment reach these global hubs first, and gradually spread to the peripheries.

The student uses topical examples, relevant concepts and manages an attempt at evaluation.

In terms of financial flows, it is clear that these are of great importance to the level of globalisation: the A.T Kearny index of globalisation: triple weights FDI flows and double weights trade volumes. FDI is the most important connection between core and periphery in terms of finance. This is because of Mydral's concept of self-sustaining growth cycles: once one industry is attracted, more will follow. FDI values are worth more than aid, and so can increase the development of often less developed peripheries. For example, China has 900 TNCs operating in Africa: connecting the core areas of China with African peripheries.

The saying 'trade not aid' illustrates the importance of the flow of goods as a financial flow between LEDCs and MEDCs.

The student links back to the question—this is made easy by using terms that appear in the question, such as "core" and "periphery".

This statement was made by Mexican coffee farmers in the 1980s, and this began the idea of Fair Trade. This is a vital flow of income and source of development in rural, LEDC peripheries. Prudent Exports and the Blue Sky organic association export Ghanaian pineapples, but invest 10% of their profits back into the farmers communities: this is connecting the core MEDC market with the Periphery producers, and allowing economic development.

China, Laos, Cambodia and Vietnam but not Thailand!

Again, the student refers to the wording of the question—this ensures the answer is focused and relevant.

Remittances are another connection between core and peripheries: family members move into a core and send their income back home to the rest of the family. This occurs in the Greater Mekong Region of Thailand, where men of working age flock from the surrounding Chinese provinces, Myanmar and Cambodia for work in a more developed region. They then send their earnings back home, linking this core with the local periphery.

Aid is another key connection between core and peripheries, in this case MEDCs and LEDCs. Project or program aid have specific development purposes in periphery areas to decrease the inequality. Examples of development aid from an MEDC core would be Japan's small scale irrigation projects in Indonesia, or their funding of HEP dams in Tunisia, where water is scarce: they are providing better water supply and electricity.

Korea is an MEDC.

Good examples—although Mexico arguably an LEDC. The answer identifies some TNCs that originate from LEDCs—this introduces an idea of change in the nature of TNCs/financial flows.

However, there are also cores in the LEDCs: export processing zones such as Incheon in South Korea, and the Maquiladoras of Tijuana in Mexico could be considered core areas due to the huge investment of TNCs and MNCs there. But these firms are not all based in MEDCs: many LEDCs now have their own MNCs, for example India's TATA, which brought the British owned Corus in 2006, and Chinese software and IT provider Newsoft, with over 80,000 employees around the world.

The student suggests alternative reasons for the lack of investment in peripheral areas, namely lack of digital connectivity, inaccessibility and political instability.

Although flows from LEDCs and MEDCs into the cores around the world are great, this investment is often slow to spread to peripheries. This could be due to the 'digital divide', inaccessibility of rural areas, political instability for example in Zimbabwe, lack of education for example the generation of child soldiers in the democratic republic of Congo and many other factors.

FDI and trade are not attracted to areas evenly or 'fairly': only peripheries with potential for resource exploitation or access to large markets, for example Mexico in NAFTA, attract the most of these forms of financial flows.

A clear attempt at evaluation of remittances.

Although aid is attracted to the most needy peripheries, it is not enough to establish a permanent two way flow: particularly in the case of tied aid or commodity aid. Remittances from members of the family who have left the periphery are often more effective than aid, as the locals can use the money for the best and most effective purpose, which may not be the case with top down development aid.

Although some peripheries remain financially unconnected, the benefits that financial flows bring to peripheries, particularly in LEDCs is huge: FDI alone is worth more than aid. [15]

Examiner report

A very good answer. The student considers both "how" and "why". The answer is focused. It is up to date with examples from a variety of locations.

The student references the question frequently and provides very good support and many evaluative comments. The response is not perfect (containing one or two errors) but well worth 15/15 for part (b).

15. Are you ready?

We hope that you enjoyed the book and that you will continue to use it throughout the year to help you to prepare for your exams.

Below is a selection of questions from recent exam papers to which you may apply your new skills. You can either take these under exam conditions or work through them, referring back to the book; the choice is yours. We have also provided the chart of markbands that examiners used when marking these papers.

Paper 1

Higher Level and Standard Level

Core Theme – Patterns and Change

SECTION A

*Answer **all** questions.*

1. Populations in transition

The graph shows the predicted population pyramid for a country in 2025.

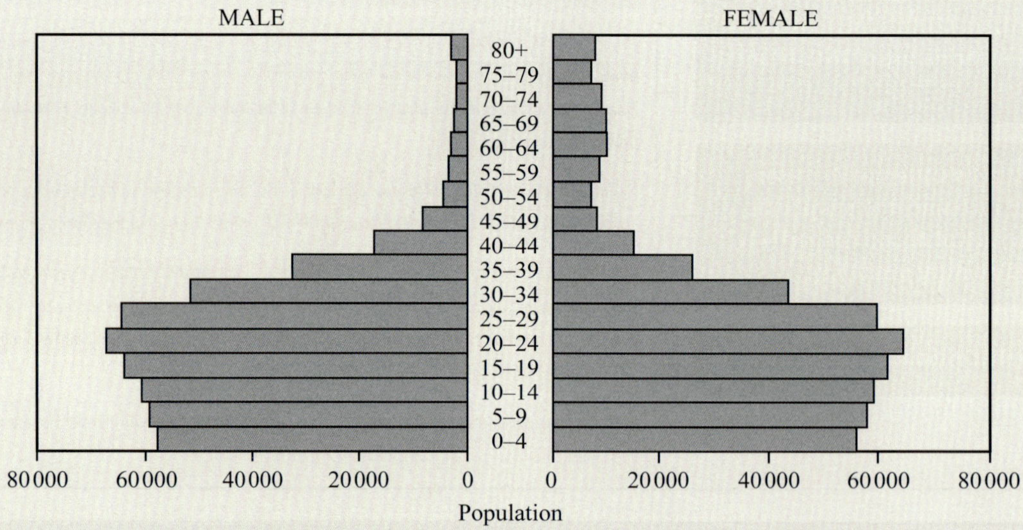

[Source: adapted from U.S. Census Bureau, International Data Base]

(a) Describe the predicted population structure of this country in 2025. *[3 marks]*

(b) Distinguish between a population projection and population momentum. *[2 marks]*

(c) Explain the economic impacts of an ageing population on a named country. *[5 marks]*

2. Disparities in wealth and development

The bar diagram shows progress towards 100% universal primary education by 2015.

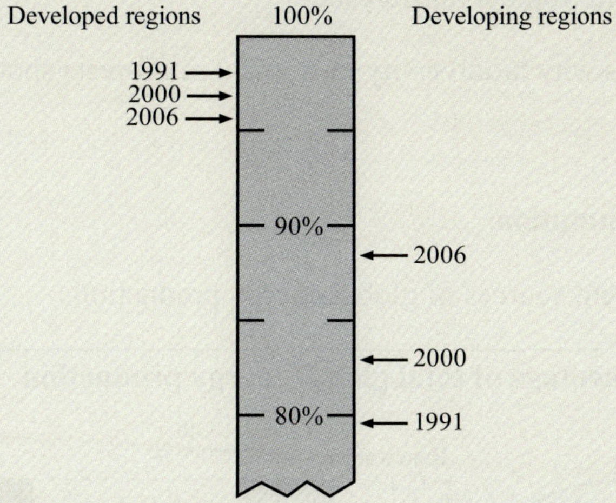

Percentage of children enrolled in primary education

[Source: MD Report, 2008]

(a) With reference to the graph, describe the regional trends in the progress made towards meeting this target. *[3 marks]*

(b) Suggest **two** reasons why developing regions have made good progress towards meeting this MDG. *[2+2 marks]*

(c) Explain how progress in education can help regions advance towards meeting **one** Millennium Development Goal (other than Universal Primary Education). *[5 marks]*

3. Patterns in environmental quality and sustainability

The diagram shows factors contributing to the loss of biodiversity in tropical rainforests.

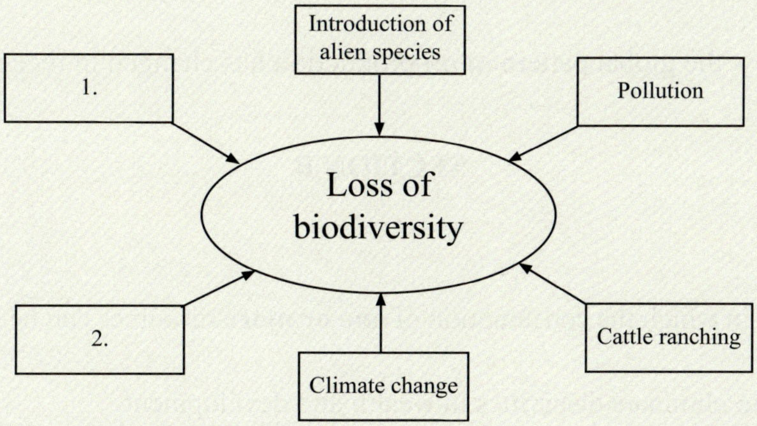

(a) State **one** physical factor and **one** human factor not shown on the diagram that contribute to a loss of biodiversity. *[2 marks]*

(Question 3 continued)

(b) Explain how any **one** of the factors **given on the diagram** contributes to a loss of biodiversity in tropical rainforests. *[4 marks]*

(c) Explain **three** reasons why biodiversity in tropical rainforests should be preserved. *[3×2 marks]*

4. Patterns in resource consumption

The graph shows the different sources of global energy production.

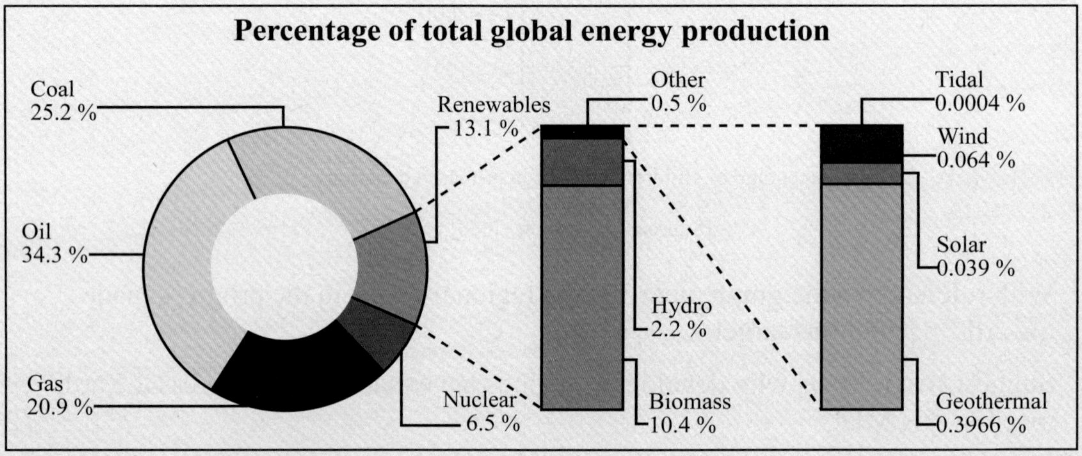

Sunlit uplands, The Economist May 31 2007, The Economist Newspaper Limited. Reproduced with permission.

(a) Identify and rank the top **three** renewable sources of energy shown on the graph from highest to lowest. *[2 marks]*

(b) Suggest **two** reasons why some areas of the world are unlikely to depend entirely on renewable energy sources. *[2×2 marks]*

(c) Analyse how the global pattern of oil production has changed in recent decades. *[5 marks]*

SECTION B

*Answer **one** question.*

5. Discuss the ways in which the consumption of **one or more** resources can be reduced. *[15 marks]*

6. "It is impossible to eliminate disparities in wealth and development." Discuss this statement. *[15 marks]*

7. Examine the view that gender inequalities are a major obstacle to development. *[15 marks]*

Paper 2

Higher Level and Standard Level

HL students answer three questions; SL students answer two questions.

Wherever possible, answers should include case studies and examples, and where relevant, large, well drawn maps and diagrams.

Optional Theme A — Freshwater – issues and conflicts

1.

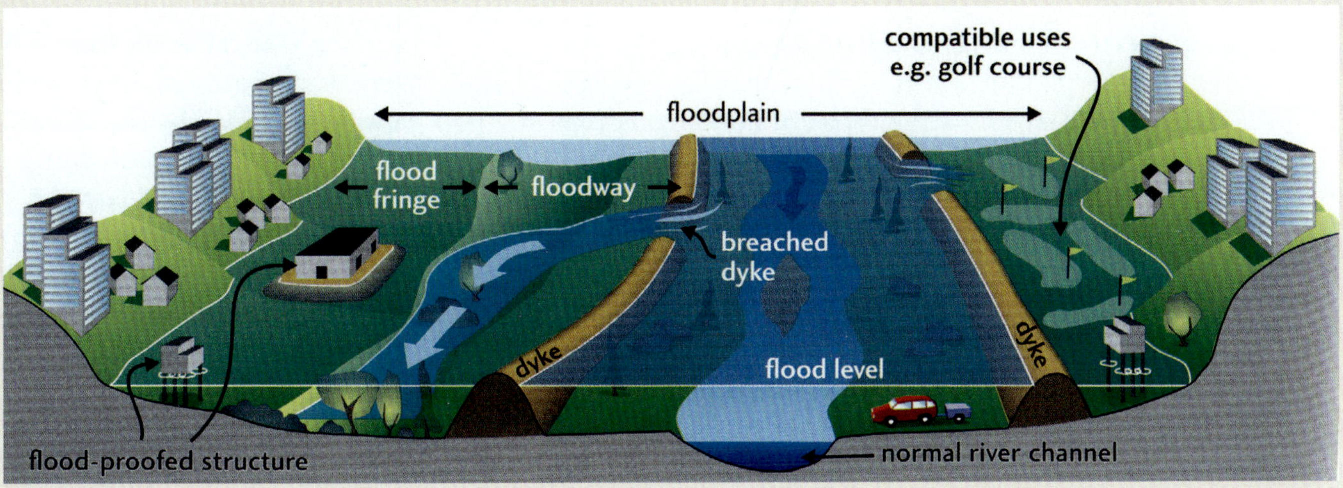

Source: John Clague and Bob Turner, (2003), *Vancouver, City on the Edge*, Tricouni Press, page 74

The diagram shows a river floodplain.

(a) Referring to the diagram, briefly describe **two** floodplain management strategies. *[2+2 marks]*

(b) Explain how human activities on a floodplain can increase the probability of flooding. *[6 marks]*

(c) "Eutrophication is the most damaging agricultural impact on water quality." Discuss this statement. *[10 marks]*

2. (a) Draw a labelled diagram of a hydrograph. *[4 marks]*

(b) Explain how hydrographs are used to forecast floods. *[6 marks]*

(c) "Dams and reservoirs create as many problems as they solve." Discuss this statement with reference to multi-purpose schemes. *[10 marks]*

Optional Theme B — Oceans and their coastal margins

3. The graph shows vertical variations in the salinity of ocean water.

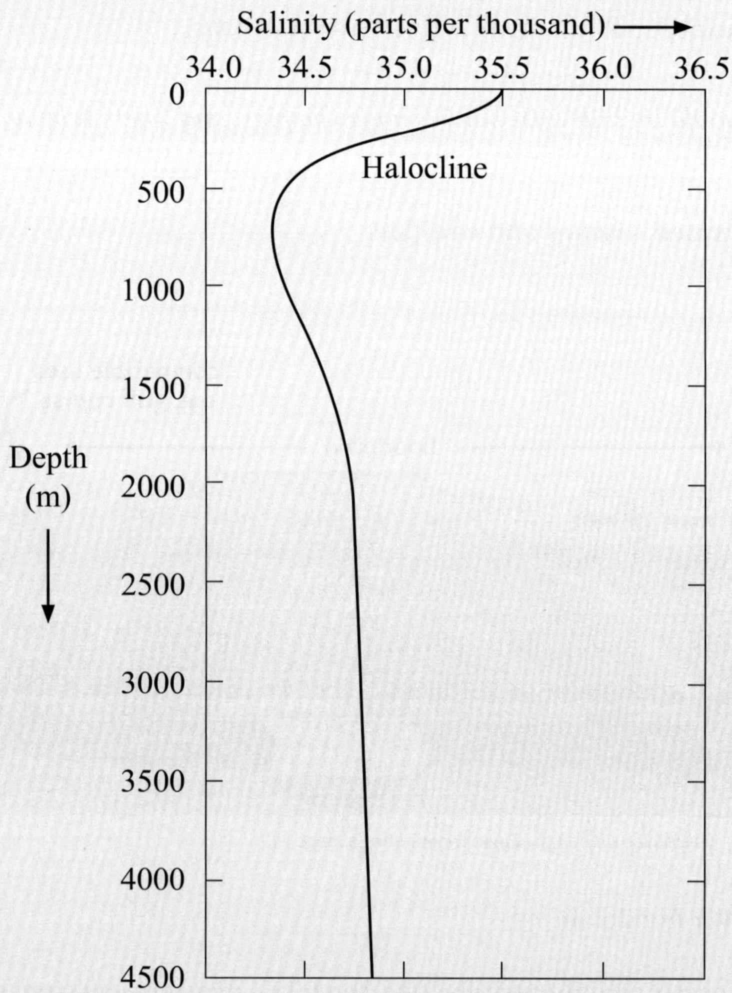

[Source: http://cimss.ssec.wisc.edu/sage/oceanography/lesson4/images/sm_salinity_depth.jpg]
University of Wisconsin-Madison Space Science and Engineering Center

(a) Describe the vertical variations in the salinity shown on the graph. *[4 marks]*

(b) (i) Explain what is meant by "oceanic conveyor belts". *[2 marks]*

(ii) Suggest **two** reasons why oceanic conveyor belts are important. *[2+2 marks]*

(c) Discuss the effects of El Niño on some coastal margins. *[10 marks]*

4.

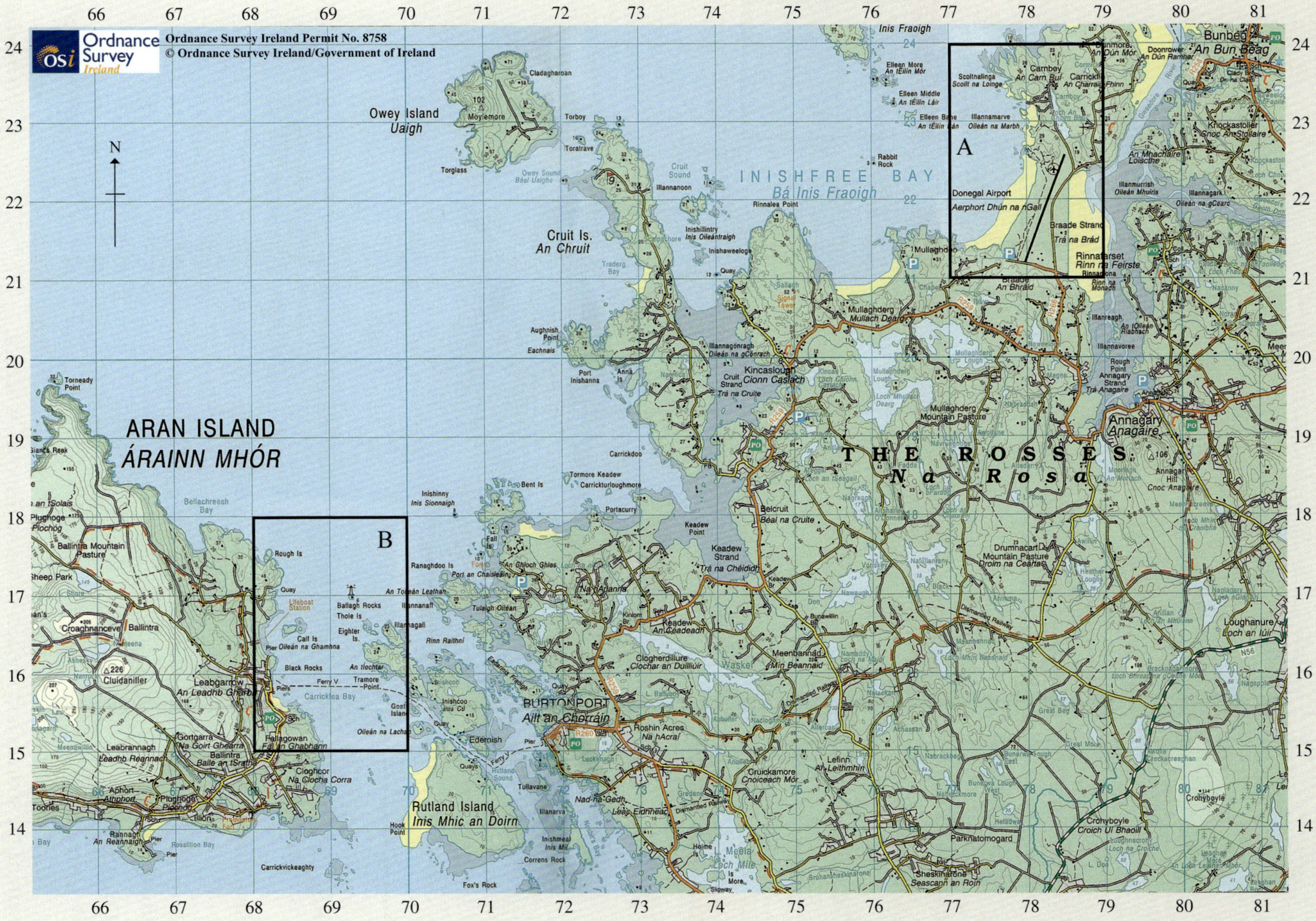

Scale 1 : 50 000

Contour interval : 50 metres

1 Kilometre 1 2 3 4 5 6 7 8 9 10

[Source: Ordnance Survey Ireland Permit No. 8758
© Ordnance Survey Ireland/Government of Ireland]

The map extract shows a coastal area in Ireland. The scale of the map is 1:50 000 and the contour interval is 50 metres.

(a) Using map evidence, describe the effects of deposition in Area A. *[4 marks]*

(b) Using map evidence, explain **one** likely conflict between types of human activity in:

(i) Area A,

(ii) Area B. *[2×3 marks]*

(c) Examine the sources and impacts of waste disposal in oceans. *[10 marks]*

Optional Theme C — Extreme environments

5.

[Source: www.coleimages.org]

The photograph shows an arid landscape.

(a) (i) Name the landform in the background of the photograph. *[1 mark]*

(ii) State **two** processes of wind transport found in areas such as this. *[2 marks]*

(iii) State the direction from which the wind normally blows. *[1 mark]*

(b) Explain the occurrence of flash floods in areas like the one shown in the photograph. *[6 marks]*

(c) "The opportunities for tourism outweigh the challenges." Discuss this statement with reference to **one** extreme environment. *[10 marks]*

6. (a) (i) Define *glacial environment.* *[2 marks]*

(ii) Define *periglacial environment.* *[2 marks]*

(b) Explain how and why glaciers retreat. *[3+3 marks]*

(c) "Periglacial areas offer more opportunities for human activity than glacial areas." Discuss this statement. *[10 marks]*

Optional Theme D — Hazards and disasters – risk assessment and response

7. The diagram shows three natural hazards (A, B, C), their duration and the size of the area they affect.

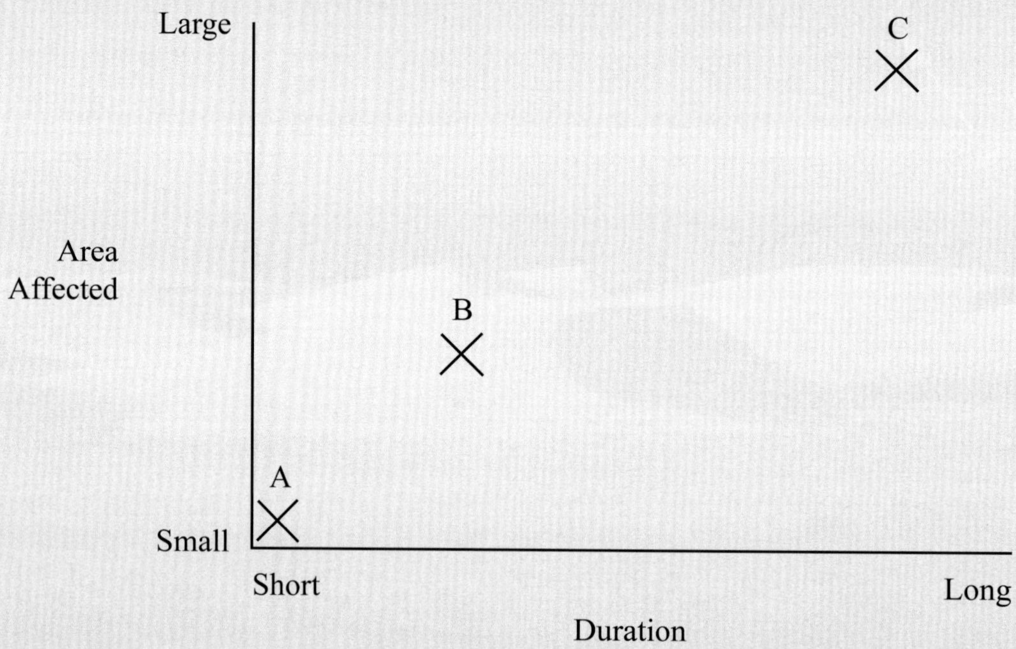

(a) Identify **two** of the hazards shown and explain your choice. *[2+2 marks]*

(b) Analyse the global distribution of **one** of the hazards you identified in part (a). *[6 marks]*

(c) Using examples, evaluate the success of adjustment and response strategies for a named hazard type. *[10 marks]*

8. (a) Outline the methods used to describe the magnitude (strength) of **two** hazard types. *[2+2 marks]*

(b) Suggest **three** reasons why some people continue to live in places with a known hazard risk. *[3×2 marks]*

(c) "Economic factors and not physical factors determine the severity of the impacts of hurricanes (typhoons, cyclones)." Discuss this statement using examples. *[10 marks]*

Optional Theme E — Leisure, sport and tourism

9. The maps show the birth places of players in a sports league in the UK in 1990 and 2010.

1990

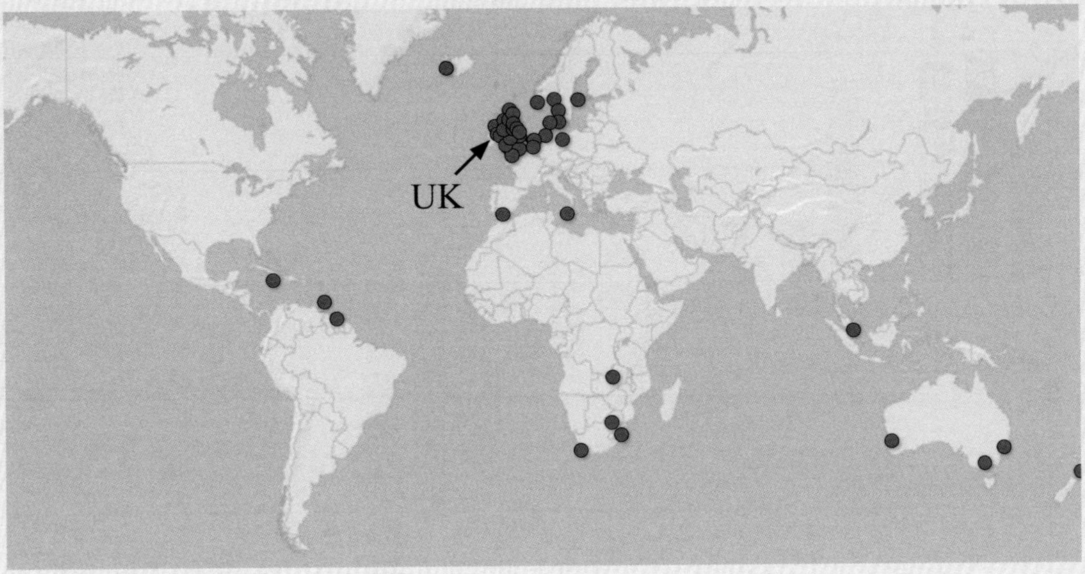

Key:
- Player's birthplace

2010

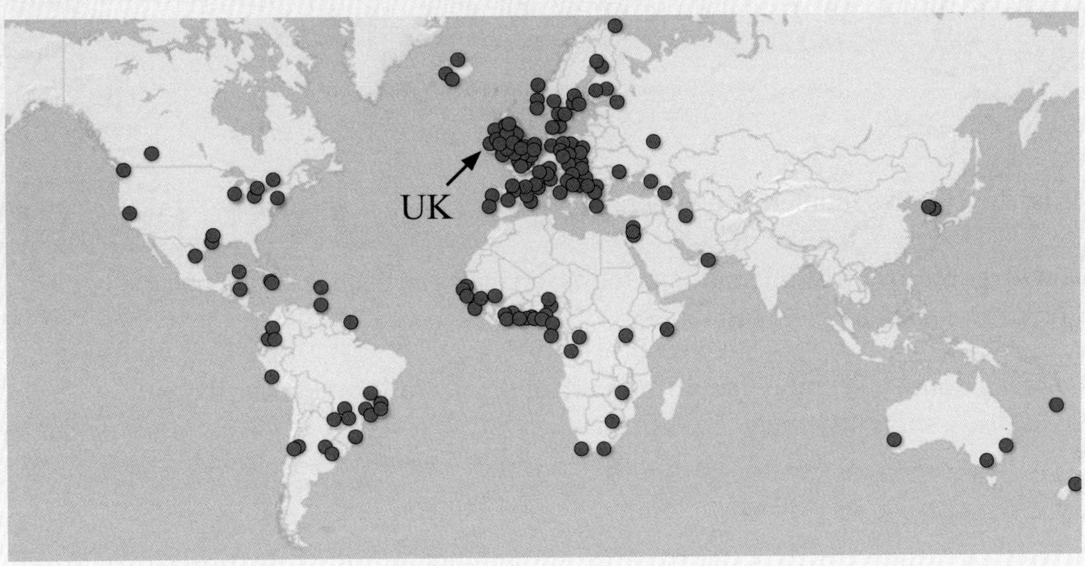

(a) Describe the pattern of birth places in 1990 and how it has changed by 2010. *[2+2 marks]*

(b) Suggest **three** possible reasons why the sphere of influence of this league's supporters has grown in size since 1990. *[3×2 marks]*

(c) "Local people do not benefit from hosting an international sporting event." Discuss this statement. *[10 marks]*

10. The graph shows the monthly changes in international tourist arrivals worldwide from January 2008 to April 2009.

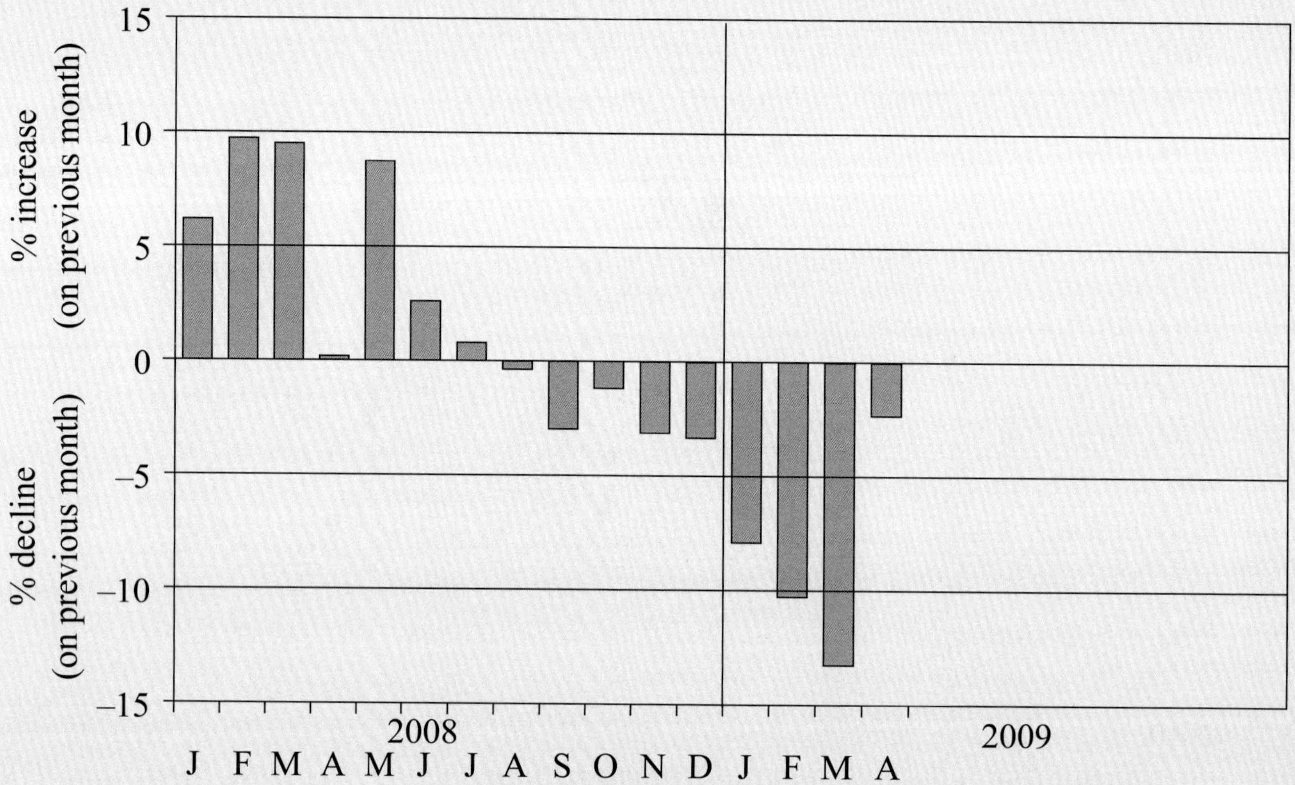

[Source: World Tourism Organization (UNWTO)]
© UNWTO, 9284404211

(a) Describe the changes in international tourist arrivals shown in the graph. *[4 marks]*

(b) Explain **three** reasons why international tourist arrivals can change in **one** named rural or urban location. *[3×2 marks]*

(c) "Most recreational and sports facilities in urban areas are located near the city centre." Discuss this statement. *[10 marks]*

Optional Theme F — The geography of food and health

11.

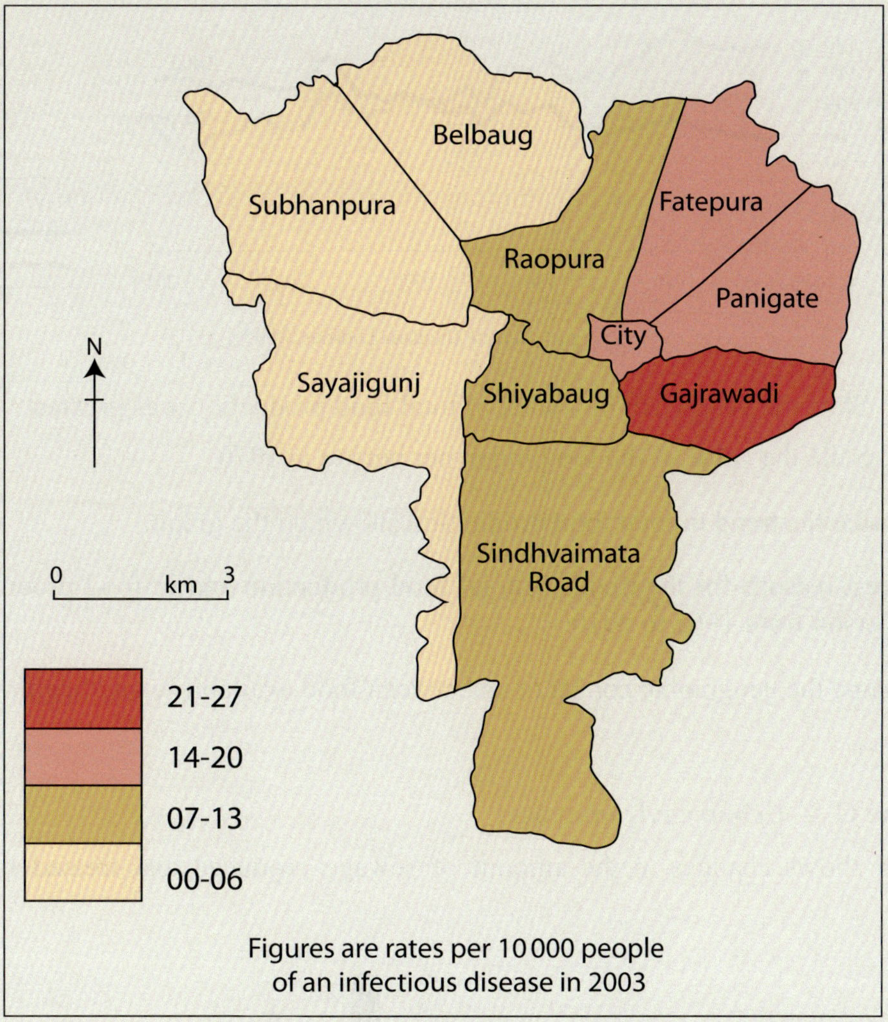

[Source: Data from Jayasree De, (Summer 2007), This Changing World: Development, Environment and Urban Health in India. *Geography*, volume 92, issue 2, pages 158–160]

The map shows the rates of infection (number of cases per 10 000 people) for a water-borne disease in a city in India in 2003.

(a) (i) State **one** other indicator besides infection rates which could be used to map the health of people in this city. *[1 mark]*

(ii) Referring to the map, describe the pattern of this disease in the city. *[3 marks]*

(b) Explain **three** possible reasons, **one** environmental, **one** social and **one** economic, for this pattern. *[3×2 marks]*

(c) "Poverty is the main cause of food insecurity." Discuss this statement. *[10 marks]*

12. The graph shows the total food production and food production per person in Country A, 1962–2010.

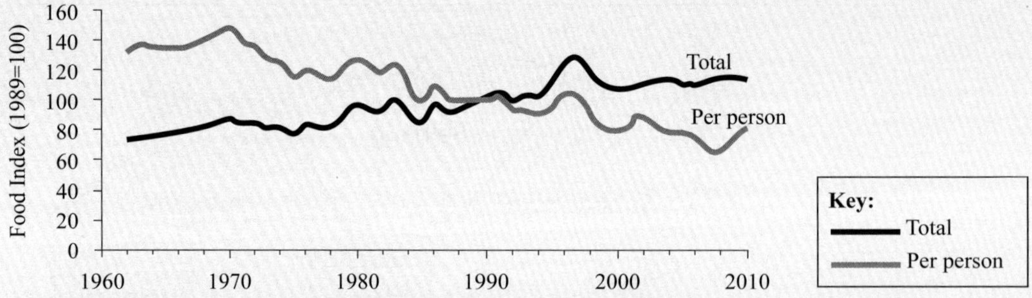

[Source: *Geofile* (Nelson Thornes, 2005)]

(a) (i) State the year when the index of total food production was greatest. *[1 mark]*

(ii) State the value of the food index per person in 1970. *[1 mark]*

(b) Describe the trend in total food production shown on the graph. *[2 marks]*

(c) Suggest reasons for the trends in total food production and in food production per person from 1962 to 2010. *[6 marks]*

(d) Examine the geographic connections between food availability and health. *[10 marks]*

Optional Theme G — Urban environments

13. The graph shows changes in the amount of sewage produced and treated in a rapidly growing megacity.

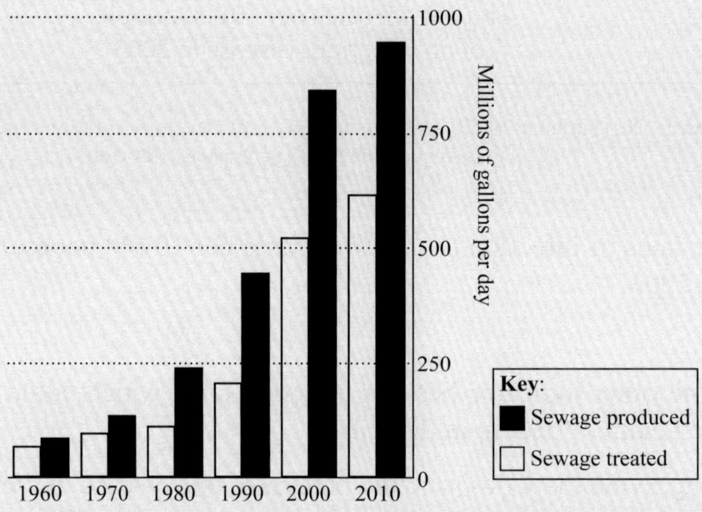

[Source: adapted from http://www.nytimes.com/imagepages/2006/09/29/world/20060929_WATER_GRAPHIC.html]

(a) Describe the trends shown in the graph. *[4 marks]*

(b) Explain **three** symptoms of urban stress. *[3×2 marks]*

(c) "Urban poverty and deprivation are widespread in all cities." Discuss this statement. *[10 marks]*

14. The diagram shows two contrasting city systems.

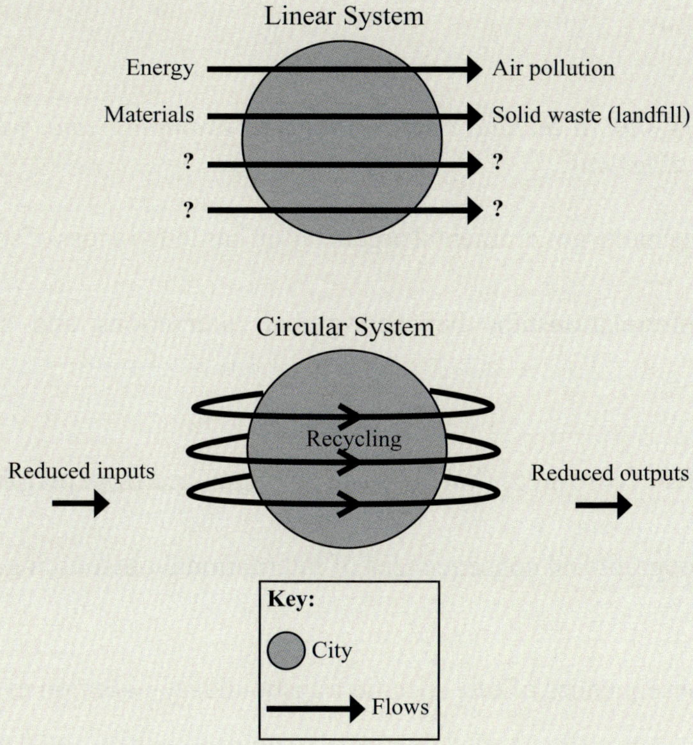

[Source: adapted from www.dep.org.uk/scities/rationale/whylearn.php]

(a) Identify **two** possible additional inputs and **two** possible additional outputs for the linear system. *[4 marks]*

(b) Explain why the circular system has a reduced urban ecological footprint. *[6 marks]*

(c) Examine the dominant population movements and their consequences for **one or more** urban areas. *[10 marks]*

Paper 3

Higher Level

*Answer **one** question.*

Wherever possible, answers should include case studies and examples, and, where relevant, large, well drawn maps, sketches and diagrams.

1. (a) Explain how **one** multi-governmental organization has led to a loss of sovereignty. *[10 marks]*

(b) Discuss the interrelationships between global interactions and changes in technology. *[15 marks]*

2. (a) Explain how global core areas (hubs) can be distinguished from peripheral areas. *[10 marks]*

(b) Examine the geographical consequences of international outsourcing. *[15 marks]*

3. (a) Analyse the consequences of **one** specific transboundary pollution event. *[10 marks]*

(b) "The negative effects of globalization on cultural traits have been overstated." Discuss this statement. *[15 marks]*

Markbands used by examiners

These markbands are devised to assist examiners in grading answers. A best-fit approach is adopted, which means that the student's answer should fulfill most, but not all, the requirements of any markband.

Paper 1 and 2 markbands

These markbands are to be used for papers 1 and 2 at both standard level and higher level.

Markband	Mark range	Mark range	Descriptor
	Paper 1 Section B	**Paper 2**	
A	0	0	AO1: No relevant knowledge; no examples or case studies AO2: No evidence of application; the question has been completely misinterpreted or omitted AO3: No evaluation AO4: No appropriate skills
B	1–3	1–2	AO1: Little knowledge and/or understanding, which is largely superficial or of marginal relevance; no or irrelevant examples and case studies AO2: Very little application; important aspects of the question are ignored AO3: No evaluation AO4: Very low level skills; little attempt at organization of material; no relevant terminology
C	4–6	3–4	AO1: Some relevant knowledge and understanding, but with some omissions; examples and case studies are included, but limited in detail AO2: Little attempt at application; answer partially addresses question AO3: No evaluation AO4: Few or no maps or diagrams, little evidence of skills or organization of material; poor terminology
D	7–9	5–6	AO1: Relevant knowledge and understanding, but with some omissions; examples and case studies are included, occasionally generalized AO2: Some attempt at application; competent answer although not fully developed, and tends to be descriptive AO3: No evaluation or unsubstantiated evaluation AO4: Basic maps or diagrams, but evidence of some skills; some indication of structure and organization of material; acceptable terminology

Markband	Mark range	Mark range	Descriptor
	Paper 1 Section B	Paper 2	
E	10–12	7–8	AO1: Generally accurate knowledge and understanding, but with some minor omissions; examples and case studies are well chosen, occasionally generalized AO2: Appropriate application; developed answer that covers most aspects of the question AO3: Beginning to show some attempt at evaluation of the issue, which may be unbalanced AO4: Acceptable maps and diagrams; appropriate structure and organization of material; generally appropriate terminology
F	13–15	9–10	AO1: Accurate, specific, well-detailed knowledge and understanding; examples and case studies are well chosen and developed AO2: Detailed application; well-developed answer that covers most or all aspects of the question AO3: Good and well-balanced attempt at evaluation AO4: Appropriate and sound maps and diagrams; well structured and organized responses; terminology sound

Paper 3 markbands

Markband	Mark range	Descriptor	Mark range	Descriptor
	Part (a) Maximum 10 marks		Part (b) Maximum 15 marks	
A	0	AO1: No relevant knowledge, or inappropriate AO2: The question has been completely misinterpreted or omitted AO4: No appropriate skills	0	AO1: No relevant knowledge, or inappropriate AO2: The question has been completely misinterpreted or omitted AO3: No synthesis/evaluation AO4: No appropriate skills
B	1–3	AO1: Little relevant knowledge and/ or understanding AO2: Important aspects of the question are ignored AO4: Little attempt at organization of material	1–4	AO1: Little relevant knowledge and/ or understanding AO2: Important aspects of the question are ignored AO3: Little attempt at synthesis/ evaluation AO4: Little attempt at organization of material

Markband	Mark range Part (a) Maximum 10 marks	Descriptor	Mark range Part (b) Maximum 15 marks	Descriptor
C	4–6	AO1: Some relevant knowledge and understanding AO2: Answer partially addresses the question AO4: Some indication of structure or organization	5–8	AO1: Some relevant knowledge and understanding AO2: Answer partially addresses the question AO3: Basic synthesis/basic or unsubstantiated evaluation AO4: Some indication of structure or organization
D	7–8	AO1: Generally accurate knowledge and understanding AO2: Answer is developed and covers most aspects of the question AO4: Appropriate structure with generally appropriate terminology	9–12	AO1: Generally accurate knowledge and understanding AO2: Answer is developed and covers most aspects of the question AO3: Synthesis that may be partially undeveloped/evaluation that may be partially unsubstantiated AO4: Appropriate structure with generally appropriate terminology
E	9–10	AO1: Accurate, relevant knowledge and understanding AO2: Well-developed answer that covers most or all aspects of the question AO4: Well structured response with sound terminology	13–15	AO1: Accurate, relevant knowledge and understanding AO2: Well-developed answer that covers most or all aspects of the question AO3: Clear, developed synthesis/clear substantiated evaluation AO4: Well-structured response with sound terminology

Source: adapted from the *Geography guide*, IBO.

Notes